U0352570

Application Theory Research of Gob-Side Entry Retaining under Specific Geology

Hongyun Yang   Yanbao Liu   Zhi Lin

Beijing
Metallurgical Industry Press
2021

内 容 提 要

本书以现场调研、理论分析、相似材料模拟、数值模拟为手段，研究煤层倾角及顶板完整性对沿空留巷适应性的影响，建立了沿空留巷适应性分级系统，介绍了特定地质条件下留巷围岩断裂破坏机理，建立了顶板控制力学模型，最后系统分析了回采巷道围岩破坏特征及控制机理。

本书可供从事矿山及地下工程的科研和工程技术人员阅读，也可供高等院校采矿专业的师生参考。

图书在版编目（CIP）数据

特定地质条件沿空留巷应用理论研究＝Application Theory Research of Gob-Side Entry Retaining under Specific Geology：英文/杨红运，刘延保，林志著．—北京：冶金工业出版社，2021.4

ISBN 978-7-5024-8850-5

Ⅰ.①特… Ⅱ.①杨… ②刘… ③林… Ⅲ.①煤矿开采—沿空巷道—研究—英文 Ⅳ.①TD263.5

中国版本图书馆 CIP 数据核字（2021）第 135461 号

出 版 人　苏长永
地　　　址　北京市东城区嵩祝院北巷 39 号　邮编　100009　电话　(010)64027926
网　　　址　www.cnmip.com.cn　电子信箱　yjcbs@cnmip.com.cn
责任编辑　刘林烨　美术编辑　吕欣童　版式设计　禹　蕊
责任校对　梅雨晴　责任印制　禹　蕊
ISBN 978-7-5024-8850-5

冶金工业出版社出版发行；各地新华书店经销；北京建宏印刷有限公司印刷
2021 年 4 月第 1 版，2021 年 4 月第 1 次印刷
710mm×1000mm　1/16；10 印张；194 千字；148 页
82.00 元

冶金工业出版社　投稿电话　(010)64027932　投稿信箱　tougao@cnmip.com.cn
冶金工业出版社营销中心　电话　(010)64044283　传真　(010)64027893
冶金工业出版社天猫旗舰店　yjgycbs.tmall.com
（本书如有印装质量问题，本社营销中心负责退换）

Preface

Coal will still be the main source of energy in China for a long time. At present, in the production and consumption structures of primary energy, coal separately accounts for 76% and 66% will occupy about 50% and 40% of China's consumption structure by 2030 and 2050, respectively. Therefore, safe, efficient, environment-friendly mining and scientific, efficient and clean utilisation of coal will be the theme driving the sustainable development of coal industry. Abiding by the idea of scientific mining and leveraging technological transformation is the main way to mine coal resources under complex and special geological conditions.

At present, with the continuous improvement of the coal mining technology, Gob-Side Entry Retaining (GER) has been widely used due to the following advantages:

(1) It changes the ventilation mode of a mining face and effectively solves the problem of gas overrun in a working face.

(2) It realizes coal mining without coal pillars, improves the rate of recovery of coal resources, and extends the service life of mines.

(3) It reduces the amount of roadway excavation and relieves the pressure of mining plans.

(4) It eliminates the isolation of the working face and barrier pillars between panels and avoids a series of safety problems caused by abutment pressure on coal pillars.

(5) It shortens the transportation time of equipment from one working face to another and prevents fire in the goaf.

(6) Gas is extracted from a pressure relief zone with GER, which realizes

co-mining of coal and gas.

(7) Gangue is used as materials for backfilling beside a roadway, which saves transportation and site occupation costs, avoiding a series of environmental problems in mines, thus realising green mining.

After years of development, the GER technology has been applied to various geological conditions, including different buried depths, roofs and floors with different lithologies and hardnesses, nearly horizontal to steeply inclined coal seams with a large dip angle, and thin, medium-thick, and thick coal seams. However, the nature of the geological conditions that allow easier implementation of GER technology, and the adaptability of GER to various geological conditions remains unclear and judgements thereof are often qualitative. Besides, a classification scheme based on its adaptability has not been formulated, which is disbeneficial to provision of targeted guidance for implementation of the GER technology.

During the development and utilisation of the GER technology, many examples show it being most widely used in mining roadways in single-structure coal seams, near-horizontal to gently inclined coal seams, thin and medium-thick coal seams, and coal seams with an intact roof. However, it is less often applied in mining roadways under specific geological conditions, such as coal seam groups, thick coal seams, steeply inclined coal seams with a large dip angle and coal seams with a weak and less intact roof. This is unfavourable for exchange and promotion of the GER technology under similar geological conditions. The reasons are shown as follows:

(1) New technologies including the GER technology developed in recent years under specific geological conditions have not been widely used and promoted. For example, at present, the technologies of GER are promoted and used by automatically forming a roadway with advanced roof cutting and natural backfilling with caved gangue in the coal seams with a large dip angle. In particularly, the technology of roadway auto-forming by roof cutting is rarely

applied in a coal seam group.

(2) Due to poor geological conditions, applying the GER technology is more difficul, and has a low success rate, such as GER in thick coal seams and the coal seams with a soft and less intact roof. Therefore, based on the background data of research, it is necessary to study and analyse the GER technique under certain specific geological conditions.

The GER needs to bear the influence of multiple excavation operations. In the early stage, due to effects of roadway excavation and primary mining, the quality of basic support during the roadway excavation and enhanced support during the primary mining exerts substantial influences on the efficacy of GER. Therefore, the quality of basic support in a roadway affects surrounding rock stability during the primary mining, while the enhanced support effects during the primary mining influence GER effects. However, at present, support parameters for excavation of most of mining roadways are designed only aiming at the early stage of utilisation, and GER in the later stage is ignored, which increases difficulties in support in the GER stage and increases support costs. Therefore, it is difficult to achieve the expected GER effects. On this basis, the needs of GER should be considered, and failure characteristics of surrounding rock of the mining roadway under changes of geological conditions, such as the dip angle, thickness, and lithology of roofs and floors of coal seams, should be better understood. Based on this, advanced and applicable supporting technologies are selected for basic support during roadway excavation and enhanced support in the process of primary mining.

Geological conditions in Sichuan Coal Industry (Group) Limited Liability Company (hereinafter referred to as Sichuan Coal Group) are diverse. On this basis, we mainly analyse the GER technique and the failure range of surrounding rock of the roadway under the specific geological conditions, such as mining roadways in near-horizontal and closely-spaced coal seams, coal seams with a large dip angle and coal seams with a soft, less intact roof. Firstly, based on the summary and analysis on a lot of GER examples, the adaptability of GER is analysed through fuzzy mathematics, and the weights of influences of different geological factors on adaptability of GER are investigated. Moreover, a classification scheme based on adaptability of GER is proposed. Secondly, through a simulation test using a similar material model, stress and evolution characteristics of deformation and failure of strata in a near-horizontal, and closely-spaced coal seam group by using the technology of roadway auto-forming by roof cutting are analysed. Furthermore, failure characteristics of strata under geological conditions of a large dip angle and a soft roof are analysed, and the key technology for controlling GER is studied. Finally, based on a ground penetrating radar (GPR) scan and numerical simulation, we analyse distributions across sections of broken rock zones (BRZ) in the mining roadway in the early stage of GER under different geological conditions. Furthermore, a technical method is developed for basic and enhanced supports. This has theoretical significance and engineering application value for safe production of coal mines, improvement of coal production efficiency, the rate of recovery of resources, and reduction of production costs in coal mines.

The authors gratefully acknowledge funding by key discipline of civil engineering of Chongqing Jiaotong University, national natural science foundation project of China (51904043), and the project of Chongqing postdoctoral innovative talents support.

Due to the limited level of the editor, there is something wrong in the book. I hope readers to criticize and correct.

The authors

January, 2021

Contents

1 Introduction ··· 1

  1.1 Classification of Mining Roadways ···························· 1

    1.1.1 Single-Index Classification Methods ···················· 1

    1.1.2 Single-Index Classification Methods Integrating Multiple Factors ········ 2

    1.1.3 Multi-Factor and Multi-Index Classification Methods ·············· 3

  1.2 Application of GER ··· 5

    1.2.1 Roadway Auto-Forming with Advanced Roof Cutting and Pressure Relief ·· 5

    1.2.2 Traditional GER Mode ·································· 5

  1.3 Failure Modes of Roadways ································· 7

  References ··· 8

2 Overview of Engineering Geology ··························· 11

  2.1 Overview ··· 11

  2.2 Coal-Bearing Strata ··· 11

    2.2.1 Coalfield in the Northeast of Sichuan Province ············ 11

    2.2.2 Huayingshan Coalfield in the East of Sichuan Province ········ 13

    2.2.3 Coalfield in the South of Sichuan Province ·············· 14

    2.2.4 Panzhihua Coalfield ···································· 15

  2.3 Distribution of Geological Conditions ···················· 15

    2.3.1 Thicknesses of Coal Seams ··························· 16

    2.3.2 Dip Angles of Coal Seams ··························· 16

    2.3.3 Burial Depths of Roadways ··························· 17

    2.3.4 Thicknesses of Immediate Roofs ····················· 17

    2.3.5 Lithologies of Immediate Roofs ······················ 17

    2.3.6 Roof Integrity ··· 18

  2.4 Chapter Summary ··· 20

  References ··· 20

3 Adaptation Assessment of Gob-Side Entry Retaining ································ 22

  3.1 Introduction ································ 22

  3.2 Determination of Impact Factors ································ 23

    3.2.1 Determination Principles ································ 23

    3.2.2 Factors Selection ································ 24

  3.3 Field Application Statistics ································ 26

    3.3.1 Dip Angle ································ 26

    3.3.2 Mining Height ································ 27

    3.3.3 Cover Depth ································ 27

    3.3.4 Lithology of Immediate Roof ································ 28

    3.3.5 TICIR ································ 28

    3.3.6 Roof Integrity ································ 30

  3.4 Indicator Weighting ································ 30

    3.4.1 Process of Determining Weights by AHP ································ 30

    3.4.2 Results of Weights ································ 32

  3.5 Adaptive Grades and Their Support Methods ································ 33

    3.5.1 The Analysis of Support Methods and Their Utilization Conditions ··· 33

    3.5.2 Adaptive Grades ································ 36

    3.5.3 Support Method of Each Grade ································ 37

  3.6 Adaptability Evaluation Process ································ 37

    3.6.1 Determination of Membership Function ································ 38

    3.6.2 Actual Gateway Adaptability Evaluation ································ 40

  3.7 Chapter Summary ································ 41

  References ································ 41

4 Advancing Cutting Roof for Retaining Gateway With Near Horizontal and Close Coal Seam Groups ································ 44

  4.1 Introduction ································ 44

  4.2 Mechanism of the Cutting Roof for a Retaining Gateway ································ 45

  4.3 Equivalent Material Simulation Experiment ································ 46

    4.3.1 Geological Conditions ································ 47

    4.3.2 Equivalent Material Model ································ 49

    4.3.3 Load and Excavation of the Model ································ 49

  4.4 Result of the Experiment ································ 50

    4.4.1 The Formation Process of the Roof Caving Structure ································ 50

4. 4. 2　The Mechanical Structure Model of the Caving Roof ······················ 53
4. 5　Discussion ··················· 59
4. 6　Chapter Summary ····················· 60
References ··················· 61

5　Natural Filling and Systematic Roof Control Technology for
　　Retaining Gateway in Steep Coal Seams ························ 64
5. 1　Introduction ····················· 64
5. 2　Investigation of Steep Coal Seams ···················· 65
5. 3　Analysis of the Advantage of the Use of GER in Steep Coal
　　　Seams ···················· 66
　　5. 3. 1　Analysis of the Characteristics of Roof Cave-Ins ···················· 66
　　5. 3. 2　Analysis of the Stress Redistribution ···················· 67
5. 4　The Key Technology for GER ···················· 71
　　5. 4. 1　Support Zone for Steep Coal Seams ···················· 71
　　5. 4. 2　Natural Filling Technology for Caved Waste Rock ···················· 72
　　5. 4. 3　Strengthening Support Device ···················· 76
5. 5　Gateway Driving and Its Support Technology ···················· 77
　　5. 5. 1　Gateway Driving ···················· 78
　　5. 5. 2　Gateway Support Technology ···················· 79
5. 6　Chapter Summary ···················· 81
References ···················· 81

6　Soft Roof Failure Mechanism and Supporting Method for Retaining
　　Gateway ···················· 84
6. 1　Introduction ···················· 84
6. 2　Description of Field Observation ···················· 85
　　6. 2. 1　Survey of Study Site ···················· 85
　　6. 2. 2　The Survey Results of Soft Roof Failure Characteristics ···················· 87
6. 3　Stress Evolution Law in Roof ···················· 88
　　6. 3. 1　Numerical Simulation Model ···················· 88
　　6. 3. 2　Numerical Results ···················· 89
6. 4　Force States of Roof Rock Mass ···················· 90
6. 5　Mechanism for Failure of Roof Rock Mass ···················· 91
　　6. 5. 1　Mechanism for Failure of Roof Rock Mass in Working Face End ······ 91
　　6. 5. 2　Mechanism for Failure of the Roof Rock Mass in Retaining
　　　　　Roadway ···················· 95

| 6. 6   Roof Support Countermeasures | 96 |

6. 6   Roof Support Countermeasures ·················· 96

6. 6. 1   Deformation Analysis of Working Face End Roof ·················· 96

6. 6. 2   Bolt Limit Equilibrium Tension Force ·················· 99

6. 7   Discussion ·················· 102

6. 8   Chapter Summary ·················· 103

References ·················· 104

7   The Failure Characteristics and the Supporting Technology for
Pre-Retaining Gateway ·················· 106

7. 1   Introduction ·················· 106

7. 2   Understanding of Surrounding Rock Failure and Its Support ·················· 107

7. 3   Investigation of Test Field ·················· 111

7. 3. 1   Geological Conditions of Gateways ·················· 111

7. 3. 2   Cross-Section and Size ·················· 112

7. 4   Test of Gateways Broken Width ·················· 113

7. 4. 1   Test Location of Gateways ·················· 113

7. 4. 2   Test Principle of GPR ·················· 113

7. 4. 3   Results and Analyses of Gateways Broken Width ·················· 115

7. 5   Cross-Section Diagram of Excavation Broken Zone (EBZ) ·················· 118

7. 5. 1   Steps for Obtaining EBZ ·················· 118

7. 5. 2   Cross-Section Diagram Resulting from EBZ ·················· 120

7. 5. 3   Failure Characteristics of Gateways Surrounding Rock ·················· 123

7. 6   Broken Width in Gateways with Different Excavation Method ·················· 123

7. 7   Experimental Studies on the Improved Material Parameters
of Bolts ·················· 125

7. 7. 1   Mechanical Effects of a Bolt with Pretension Force ·················· 125

7. 7. 2   The Improved Mechanical Performance of a Bolt ·················· 127

7. 8   Support Effect Analyses of Bolt and Surround Rock ·················· 131

7. 8. 1   Support Theory Assessment of Gateways Broken Zones ·················· 131

7. 8. 2   Stress Diffusion Analysis for Anchored Rock Mass ·················· 132

7. 8. 3   Initial load-Bearing Zone of Gateway Surrounding Rock ·················· 134

7. 9   Supporting Technology ·················· 137

7. 10   Chapter Summary ·················· 138

References ·················· 140

Attachment   Evaluation Results of Adaptability of Gob-side Entry
Retaining ·················· 144

1 Introduction

1.1 Classification of Mining Roadways

The classification of mining roadways mainly refers to classification of surrounding rock stability, that is, the surrounding rock stability of roadways is classified according to the difficulties of support, to provide a scientific basis for design, construction, and management of roadway support. The current classification methods of surrounding rock include single-index classification methods, single index classification methods integrating multiple factors, and multi-factor and multi-index classification methods[1, 2].

1.1.1 Single-Index Classification Methods

The typical single-index classification methods include Protodyakonov classification, rock quality designation (RQD) and comprehensive classification based on the elastic wave velocity of rock mass. These methods do not consider the effect of *in-situ* stress, so in recent years, their sole use has not been advocated: they are taken as an index for other classification methods of surrounding rock.

1.1.1.1 The Protodyakonov Classification

The Protodyakonov classificationwas proposed by Russian scholars in 1907. For this method, one tenth of uniaxial compressive strength R_c of rock is taken to define the solidity coefficient of rock (namely the Protodyakonov coefficient) f, that is, $f = R_c/10$. The surrounding rock of the roadway is classified into 15 types including the strongest, stronger, and strong rock.

1.1.1.2 The RQD-Based Classification

The RQD-based classification was proposed by scholars in the United States in 1967. Firstly, it is necessary to drill rock cores and then to calculate the cumulative length of all rock cores (length is more than 100mm). The percentage ratio of the cumulative length to the length of a drill-hole is the RQD value.

1. 1. 1. 3   The Comprehensive Method

The comprehensive method based on the elastic wave velocity of rock mass was proposed by scholars in Japan. Based on the sensitivity of the elastic wave velocity to joint fractures, the integrity of rock mass is evaluated and the integrity coefficient is $K_v = (v_p/v_p^0)^2$, where v_p and v_p^0 separately represent the acoustic velocities of rock mass and rock blocks.

1. 1. 2   Single-Index Classification Methods Integrating Multiple Factors

The methods include RQD-based classification (Q system), rock mass rating (RMR), coal mine roof rating (CMRR)[3, 4], rock mass quality rating (RMQR)[5, 6], classification of surrounding rock based on deformation, and classification of surrounding rock based on BRZs.

1. 1. 2. 1   The Q System

The Q system was proposed by Barton, working in Norway in 1974. This systematic classification method reflects the effects of important factors including joints and *in-situ* stress of surrounding rock, which makes the classification more scientific. However, because many factors are involved, the quantitative determination of indices is arbitrary, thus limiting its practical application.

1. 1. 2. 2   The RMR Method

The RMR method was developed by Bieniawski in 1973 and has been modified and improved over 16 years[7]. The RMR consists of five groups of indices. From the point of view of field application, it is easy to use and many RMR indices are useful. But the method is inapplicable to extremely weak rock mass, such as rock mass subjected to extrusion, expansion and water gushing.

1. 1. 2. 3   The CMRR Method

The CMRR method was proposed by Molinda and Mark in the United States in 1994. It includes four main parameters (that is, uniaxial compressive strength of intact rock mass, distribution parameters of discontinuities, shear strength of joint surfaces, and humidity sensitivity) and other secondary parameters, with a value range of 0 to 100. The roof rock is divided into three grades, namely, an unstable roof (CMRR< 45), a moderately stable roof (45≤CMRR≤65), and a stable roof (CMRR>65).

1. 1. 2. 4   The RMQR Method

The RMQR method was proposed by Japanese scholar Aydan in 2013. The RMQR includes six parameters with values ranging from 0 to 100. The rock mass quality is ranked into six grades, as shown in Table 1. 1.

Table 1. 1   Grades of rock mass quality according to RMQR values

Grade of rock mass	I	II	III	IV	V	VI
Characteristic	Intact	Very good	Good	Medium	Poor	Very poor
RMQR	100≥ RMQR>95	95≥ RMQR>80	80≥ RMQR>60	60≥ RMQR>40	40≥ RMQR>20	20≥ RMQR

1. 1. 2. 5   Classification of Surrounding Rock Based on Deformation

This method was proposed based on the understanding that the deformation of surrounding rock is the result of many factors after the roadway excavation. This method classifies surrounding rock stability into five classifications according to deformation of surrounding rock. It is simple and convenient to use in engineering practice.

1. 1. 2. 6   Classification of Surrounding Rock Based on BRZs

The size of the BRZs in surrounding rock is the result of interaction of multiple factors including stress, property of rock mass, construction and water. The parameter of the size of the BRZs is quantified and has a clear physical meaning, so it is widely used. However, an explicit analytical formula for the relationship between the size of the BRZs in surrounding rock and the main influencing factors is not included.

1. 1. 3   Multi-Factor and Multi-Index Classification Methods

With the constant development of mathematical theory and computer technology, these, such as fuzzy mathematics, artificial neural network and grey theory, have been applied in the classification of surrounding rock stability of the roadway. They promote the rapid development of classification of surrounding rock stability and improve classification accuracy.

1. 1. 3. 1   Classification Based on the Theory of Fuzzy Mathematics

This theory solves the contradiction between accuracy and complexity and it can be used to study the phenomena and objects with unclear boundaries and fuzziness. The main methods applied to the classification of surrounding rock stability of the roadway including the fuzzy clustering analysis, fuzzy comprehensive evaluation, and fuzzy identification.

Seven classification indices, including surrounding rock strength of the roadway, width of a coal pillar and integrity index of rock mass, have been delineated in the *Classification of Surrounding Rock Stability of a Mining Roadway with Gently Inclined and Inclined Coal Seams in China* published in 1988. By using fuzzy clustering analysis, surrounding rock stability of the mining roadways is classified into five types[8]. By combining with the specific geological conditions of the roadway in the studied mining areas, scholars classified surrounding rock stability of the mining roadways by selecting a certain evaluation index using fuzzy clustering (or fuzzy comprehensive evaluation and fuzzy identification) [9-12].

1.1.3.2　Classification Based on an Artificial Neural Network

Through different learning and training, the weights are adjusted dynamically, so that the neural network has the intelligent computing abilities, such as classification and prediction. Therefore, the artificial neural network is applied to identify the stability of the rock surrounding a roadway.

1.1.3.3　Classification Based on the Grey Theory

Among factors affecting surrounding rock stability of a roadway, some are clear, others are not, thus invoking the power of a grey system classifier.

There are many factors affect stability of the rock surrounding a roadway, and each factor represents a certain attribute thereof. Selecting more main factors as the classification indices for surrounding rock stability can provide more comprehensive evaluation results. As the adopted theories and techniques improve, the multi-index classification method offers very high mapping accuracy and especially. The combined application of multiple methods can make the most of strong points and avoid weak points, thus improving classification accuracy[13]. In addition, although there are many classification methods for mining roadways, they all aim to support surrounding rock during roadway excavation, while the classification of mining roadways using GER is not involved. Therefore, it is necessary to grade and study difficulties of GER under geological conditions of the rock surrounding each roadway by utilising the multi-index classification method.

1.2 Application of GER

It can be found from the development of the GER technique that there are two modes including GER without backfilling beside a roadway, and GER by backfilling beside a roadway. The former involves roadway auto-forming with advanced roof cutting and pressure relief developed mainly in recent years, and the latter is a mode of GER that has been used all the time (the traditional GER mode).

1.2.1 Roadway Auto-Forming with Advanced Roof Cutting and Pressure Relief

The method of roadway auto-forming by roof cutting and pressure relief is a technical revolution in coal mining. In this technology, blastholes for pre-splitting blasting are arranged on the side near the working face in the reserved roadway in front of the working face, to conduct advanced pre-splitting blasting on the roof of the working face, so that the roof produces cutting slots along the predetermined direction. As the working face constantly advances, the roof in the goaf is subjected to periodic weighting, and the roof of the roadway near the goaf collapses along the cutting slots. The cut roof forms a new side wall to support the overlying strata and isolate the goaf. Meanwhile, the roof of the roadway is supported by constant-resistance anchor bolts and cables, and the support is reinforced with single hydraulic props. This technique was first applied by He Manchao[14] to the hard roof of the transportation roadway of the No. 2422 working face in protective seams of Baijiao Coal Mine of Sichuan Coal Group Furong Company (hereinafter referred to as Furong Company) in 2010.

1.2.2 Traditional GER Mode

The traditional GER by backfilling beside a roadway mainly involves a mechanical model of roof movement, the mode of backfilling beside a roadway, and lithologic conditions of the roof based on the mechanical model.

1.2.2.1 The Mechanical Model of Roof Movement

At the beginning of the design of backfill materials beside an artificial roadway with GER, it is necessary to assess the movement of the rock surrounding a roadway and characteristics of strata behaviours, and to establish the mechanical model of roof movement.

Whittaker et al. in the United Kingdom proposed a mechanical model of separated rock blocks, and Smart et al. from the University of South Wales in the United Kingdom proposed a mechanical model of roof inclination.

In the early stage, Chinese scholar Sun Henghu simplified the roof during GER into a rectangular superimposed laminate structure with negligible interlayer bonding force in 1993. In about 2000, several mechanical models of roofs appeared[15]. After 2010, scholars developed a new understanding of the forms of movement of the rock mass in the roadway roof, thus establishing many new mechanical models of roof movement during GER[16-21].

1.2.2.2　Backfilling Beside the Roadway

According to the mechanical characteristics of movement of roof strata, backfilling beside the roadway has, as a technique, been constantly improved to adapt to roof subsidence and dislocation. In Germany, low-water-content materials, such as anhydrite, fly ash, Portland cement, gangue, and cementitious materials, were used for backfilling beside the roadway. After that, this method was developed in the United Kingdom and Poland. In 1979, the method of backfilling with high-water-content materials beside the roadway was tested underground in the United Kingdom and developed rapidly thereafter. It once accounted for about 90% of the total number of backfilling operations besides the roadway and the method was also used in Germany.

Backfilling beside the roadway in China has developed from backfilling with wood piles, gangue walls (bags), dense pillars, and concrete blocks to high-strength paste, high-water-content materials and concrete materials. It has been widely used in near-horizontal, gently inclined, and thin and medium-thick coal seams[22]. With the emergence of the support technique using high-strength and high-preload anchor bolts and cables, combined with the role of backfilling beside the roadway, GER has achieved remarkable results. In recent years, with the development of the GER technique, the mode of GER by natural backfilling caved gangue beside a roadway has begun to be used in the coal seams with a large dip angle[23-25].

1.2.2.3　Lithologic Conditions of the Roof

Cao Shugang[26] statistically analysed the lithology of the immediate roofs of 136 roadways through GER, of which there are 49 roadways with soft roofs containing sandy mudstone, argillaceous sandstone, and mudstone, accounting for 36%. It is found that many existing mechanical models are applied to the roadways with a hard and stable or a moderately stable roof, but not suitable for the roadways with soft roofs. GER is more difficult to apply to soft roofs compared with hard and stable, or moderately stable roofs[27-30].

1.3 Failure Modes of Roadways   · 7 ·

For the two GER modes, basic and enhanced supports are used in the roadway. In the early stage, the basic support modes mainly include supports with I-steel beams and U-shaped yieldable steel ribs. However, under some complex geological conditions, the early basic support is still used in the mining roadway with the closely-spaced coal seam group. At present, anchor bolts, anchor cables, and combined support with anchor (cables) and mesh belts (beams) have been mainly used for basic support. Under complex geological conditions, a support system with high-preload and strong anchor bolts and cables is developed. In many mining areas, high-preload and strong anchor bolts and cables create conditions for support in the roadway, and enhanced support with anchor cables beside the roadway during GER, which provides basic support to the roadway during its formation by roof cutting. Furthermore, for the enhanced support, the single hydraulic props combined with hingedbar support and a small number of high-resistance hydraulic supports specially designed for GER are generally used.

In conclusion, the technique of roadway auto-forming by roof cutting without backfill materials has been effectively utilised in the environment of single coal seams and simple geological conditions. However, under complex occurrence conditions of the coal seam group, GER mechanism warrants further exploration. Backfilling beside the roadway requires a development process from artificial backfilling in near-horizontal and gently inclined coal seams to natural backfilling with caved gangue in coal seams with a large dip angle, while the later warrants further study. Besides, it is more difficult to use GER in the roadway with a soft roof and there are few mechanical models available as a guide. Therefore, it is necessary to analyse failure mechanisms of the rock surrounding a roadway with a soft roof and countermeasures for roadway support under the conditions in the presence of backfilling materials.

1.3 Failure Modes of Roadways

Protodyakonov theory (1907) and Terzaghi's theory (1942) consider that a caving arch can be automatically formed after the excavation of underground caverns and the force acting on the support structure is the weight of rock within the range of the caving arch; The former holds that the caving arch is a parabolic arch, while the latter treats it as a rectangle. With the development of elasto-plastic theory, based on the M-C, H-B, and D-P criteria, the Fenner formula and Kastner formula are obtained through the classical ideal elasto-plastic model and the assumption of unchanged volume of rock after failure. The distribution range of a plastic failure zone of surrounding rock of the circular cavern can be calculated according to different stress and lithological conditions of surrounding rock, and failure modes of the surrounding rock are determined.

In the 1950s, Yu Xuefu assumed that rock mass is a homogeneous and isotropic elastic body and proposed the theory of axial variation for the deformation and failure of surrounding rock of a roadway. In the 1990s, Dong Fangting proposed the theory of BRZs of surrounding rock. With the development of experimental methods, especially those using simulation of similar materials, the failure modes of the rock surrounding a roadway have been analysed using finite element, discrete element, finite difference theory and advanced test techniques. Xin Yajun[31] analysed the failure characteristics of the rock surrounding a special-shaped mining roadway in soft rock with a large dip angle. Gou Panfeng[32] studied stability of a roadway under different horizontal stress conditions through a similar material model and a numerical simulation experiment. Zhao Zhiqiang[33] derived the boundary equation of a plastic zone in the rock surrounding a circular roadway under non-uniform stress. The numerical simulation shows that key factors, such as a butterfly-shaped irregular plastic zone and its mechanical behaviour, magnitude and direction of stress field, shapes of cross-sections of the roadway and lithological combination, exert different influences on morphology of the plastic zone. Yu Yuanxiang[34] investigated the failure modes and ranges of the roof of a rectangular roadway and considered that failure modes on both sides of the roadway are related to the lateral pressure. Kang Qinrong[35] tested and analysed the thickness of a BRZ during cross-mining of an air return uphill roadway in a mining area by using GPR scanning.

To sum up, there are many research results on failure ranges and modes of rock surrounding mining roadways, which provides a reference for the selection of support parameters. However, these studies are mainly based on the theory of ideal elasto-plastic mechanics under specific geological conditions and continuous media. Some scholars have tested BRZs under the specific geological conditions in the field, and more systematic and in-depth research are required. In fact, the prevailing geological conditions around mining roadways in different mining areas differ greatly, and the failure of the rock surrounding roadways under various conditions warrant analysis.

References

[1] Liu Q S, Gao W, Yuan L. Stability Control Theory and Support Technology and Application of Deep Rock Roadway in Coal Mine[M]. Beijing: Science Press, 2010.

[2] Cai M F, He M C, Liu D Y. Rock Mechanics and Engineering[M]. Beijing: Science Press, 2002.

[3] Molinda G M M, Christopher. The Coal Mine Roof Rating (CMRR): A Practical Rock Mass Classification for Coal Mines[M]. Washington, D. C: United States Department of the Interior.

Bureau of Mines, 1994.

[4] Mark C, Molinda G M. The Coal Mine Roof Rating (CMRR)——A decade of experience[J]. Int J Coal Geol, 2005, 64(1,2): 85-103.

[5] Aydan Ö, Ulusay R, Tokashiki N. A new rock mass quality rating system: Rock Mass Quality Rating (RMQR) and its application to the estimation of geomechanical characteristics of rock masses[J]. Rock Mech Rock Eng, 2014, 47(4): 1255-1276.

[6] Aydan Ö, Ulusay R, Tokashiki N. Rock Mass Quality Rating (RMQR) system and its application to the estimation of geomechanical characteristics of rock Masses [J]. Engineering Geology for Society and Territory——Volume 6: Applied Geology for Major Engineering Projects. Cham, Springer International Publishing. 2015: 769-772.

[7] Aksoy C O. Review of rock mass rating classification: Historical developments, applications, and restrictions[J]. J Min Sci, 2008, 44(1): 51-63.

[8] Liu Y T, Hou C J, Yao J G , et al. The classification scheme of surrounding rock stability of mining roadway in gently inclined and inclined coal seams in China[J]. Coal Sci Technol, 1988, 6: 2-6,61-62.

[9] Li S G, Zhang K B, Zhang K Z. Study on the fuzzy equivalence cluster of stability of surrounding rock of gateways in coalmines[J]. J Shandong U Sci Technol, 2003, 22(4): 18-20.

[10] Li H, Jiang J Q, Zhang K Z. Fuzzy clustering analysis of tunnel surrounding rock classification [J]. J Xi' an U Sci Technol, 2005, 25(1): 12-16.

[11] Li Y F. Stability classification and control technique of surrounding rock in deep mining roadway [J]. Safety Coal Min, 2013, 44(6): 75-78.

[12] Yang R S, Wang M Y, Ma X M. Research on surrounding rock stability classification of coal drift[J]. Coal Sci Technol, 2015, 43(10): 40-45,92.

[13] Kong X S. Study on coal roadway surrounding rock stability classification and strong sidewall and corner supporting technology in Shanxi coking coal group[D]. Beijing: China University of Mining and Technology, 2014.

[14] He M C. Research on the key technology of no coal pillar mining and gob-side entry retaining with cutting roof for pressure relief [D]. Sichuan: Sichuan Furong Group Industrial Co. LTD, 2010.

[15] Kang H P, Niu D L, Zhang Z. Deformation characteristics of surrounding rock and supporting technology of gob-side entry retaining in deep coal mine[J]. Chinese J Rock Mech Eng, 2010, 29(10): 1977-1987.

[16] Chen Y. Study on stability mechanism of rockmass structure movement and its control in gob-side entry retaining[D]. Xuzhou: China university of mining and technology, 2012.

[17] Tan Y L, Yu F H, Ning J G, et al. Design and construction of entry retaining wall along a gob side under hard roof stratum[J]. Int J Rock Mech Min, 2015, 77:115-121.

[18] TanY L, Yu F H, Ning J G, et al. Adaptability theory of roadside support in gob-side entry retaining and its supporting design[J]. J China Coal Soc, 2016, 41(2): 376-382.

[19] Han C L, Zhang N, Li B Y, et al. Pressure relief and structure stability mechanism of hard roof

for gob-side entry retaining[J]. J Cent South Univ, 2015, 22(11): 4445-4455.

[20] Feng X, Zhang N. Position-optimization on retained entry and backfilling wall in gob-side entry retaining techniques[J]. Int J Coal Sci Technol, 2015, 2(3): 186-195.

[21] Kan J G, Zhang N, Li B Y, et al. Analysis of supporting resistance of backfilling wall for gob-side entry retaining under typical roof conditions [J]. Rock Soil Mech, 2011, 32(9): 2778-2784.

[22] Guo Z, Mou W, Huang W, et al. Analysis on Roadside Support Method with Constant Resistance Yielding-Supporting Along the Goaf Under Hard Rocks[J]. Geotech Geol Eng, 2016, 34(3): 827-834.

[23] Zhou B J. Research on compatible deformation mechanism between backfill body-surrounding rock and gob-side entry retaining technology[D]. Xuzhou: China University of Mining and Technology, 2012.

[24] Zhou B, Xu J, Zhao M, et al. Stability study on naturally filling body in gob-side entry retaining [J]. Int J Min Sci Technol, 2012, 22(3): 423-427.

[25] Deng Y, Wang S. Feasibility analysis of gob-side entry retaining on a working face in a steep coal seam[J]. Int J Min Sci Technol, 2014, 24(4): 499-503.

[26] Cao S G, Chen X Z, Yang H Y, et al. Analysis on roadside control technology of gob-side entry retaining and applicable conditions[J]. Coal Sci Technol, 2016, 44(4): 27-33.

[27] Dai J, Qiang H L, Liu L L, et al. Numerical analysis of different immediate roof influenced by key piece[J]. Coal Technol, 2015, 34(4): 4-6.

[28] Li Z H, Hua X Z, Li Y F. Deformation characteristics of surrounding rock in deep mine gob-side entry retaining with different strength of roof and floor[J]. Coal Eng, 2016, 48(5): 91-93,97.

[29] Ju F, Sun Q, Huang P, et al. Study on technology of gob-side entry retaining in thin seam surrounded by soft roof and floor[J]. J Min Safety Eng, 2014, 31(6): 914-919.

[30] Zou D J. Research of maching technolgy for gob-side entry retaining in the thin coal seams group with vevy short distance and "Three Soft" [D]. Chongqing: Chongqing University, 2013.

[31] Xin Y J, Gou P F, Yun D F, et al. Instability characteristics and support analysis on surrounding rock of soft rock gateway in high-pitched seam[J]. J Min Safety Eng, 2012, 29(5): 637-643.

[32] Gou P F, Wei S J, Zhang S. Numerical simulation of effect of horizontal stresses at different levels on stability of roadways[J]. J Min Safety Eng, 2010, 27(2): 143-148.

[33] Zhao Z Q. Mechanism of surrounding rock deformation and failure and control method research in large deformation mining roadway[D]. Beijing: China University of Mining and Technology, 2014.

[34] Yu Y X. Study on the deformation mechanism of the surrounding rock in rectangular laneway and its application in wangcun coal mine[D]. Xi'an : Xi'an University of Science and Technology, 2013.

[35] Kang Q R. Research of surrounding rock stability control of floor space soft rock roadway in gently inclined coal seams group[D]. Chongqing: Chongqing University, 2011.

2 Overview of Engineering Geology

2. 1 Overview

The diversity of coal seams can provide rich geological resources for technological and theoretical research into the application of GER. Coal-bearing strata in the coal mine of Sichuan Coal Group are controlled by the paleogeographic environment and changes during deposition. The occurrence conditions of coal seams, such as thickness and dip angle of coal seams as well as roof lithology and integrity are quite different, and the geological resources are diversified. Based on the coal-bearing strata of a coal mine subordinating to Sichuan Coal Group, here we analyse the engineering geological conditions and characteristics in detail.

2. 2 Coal-Bearing Strata

The coalfields of Sichuan Coal Group are mainly distributed around the Sichuan Basin and Panzhihua area. Sichuan Basin lies in the Yangtze Plate, with the Qinling-Micangshan-Dabashan nappe orogenic belt in the north. It is a multicycle sedimentary basin, which was shaped after the transformation from Yanshanian movement to Himalayan movement. There are coalfields in the north-east of Sichuan Province, Huayingshan Coalfield and a coalfield in the south of Sichuan Province around the basin. The coal seams of the Xujiahe Formation in Triassic and Longtan Formation and Xuanwei Formation in Permian are mainly mined. Panzhihua Coalfield is mainly located in the Panzhihua area in the south of Sichuan Province and coal seams in the Daqiaodi Formation in Permian are mainly mined. These coalfields are the main mining areas of Sichuan Coal Group. Some formations in the Permian and Triassic strata in each coalfield are listed in Table 2. 1.

2. 2. 1 Coalfield in the Northeast of Sichuan Province

The coalfield in the north-east of Sichuan Province is mined by Guangwang and Dazhu Companies, and the Triassic Xujiahe Formation is the main coal-bearing strata with its thickness of about 400m to 2200m[1], showing sedimentary characteristics of the foreland basin. The strata mainly include clastic rock, such as conglomerate, sandstone,

Table 2.1　Coal bearing strata of Sichuan coal industry group

Facies Formation	Mining area					
	Panzhihua mining area	Mining area in the south of Sichuan Province		Huayingshan mining area (Mining area in the east of Sichuan Province)	Mining area in the north-east of Sichuan Province	
		Furong mining area	Guxu mining area		Guangwang mining area	Dazhu mining area
	Continental facies	Continental facies	Marine and continental alternative facies	Marine and continental transitional facies	Continental facies	
Triassic — Upper series	Baoding Formation: Upper member / Lower member; Daqiaodi Formation: Multiple members	Xujiahe Formation	Xujiahe Formation	Xujiahe Formation	Xujiahe Formation: Member 7 / Member 6 / Member 5 / Member 4 / Member 3 / Member 2 / Member 1	Xujiahe Formation
Triassic — Middle series	Leikoupo Formation					
Triassic — Lower series	Bingnan Formation	Jialingjiang Formation / Tongjiezi Formation	Jialingjiang Formation / Tongjiezi Formation	Jialingjiang Formation	Jialingjiang Formation	Jialingjiang Formation
Triassic — Lower series	Feixianguan Formation					
Permian — Upper series	Emeishan Basalt Formation	Changxing Stage / Wujiaping Stage	Xuanwei Formation: Upper section / Middle member / Lower member	Changxing Formation; Longtan Formation: Upper member / Lower member	Changxing Formation; Longtan Formation: Member 5 / Member 4 / Member 3 / Member 2 / Member 1	Dalong Formation / Wujiaping Formation
Permian — Upper series	Emeishan Basalt Formation					
Permian — Middle series	Maokou Formation					

siltstone, and mudstone, and are intercalated with coal seams and a small amount of siderite. In the strata, Guangwang Coalfield mainly appears as the front edge of an alluvial fan, while the coalfields to the east of Luzhou and Kaijiang are shown as alluvial riverine plains. The coal-bearing strata are divided into seven sections from bottom to top, in which Members 2, 4, and 6 are coarse clastic rock, while Members 1, 3, 5, and 7 consist mainly of fine-grained sediments. In the north-west of the basin, Member 6 is denuded, and Member 7 is only developed in the central and eastern parts thereof[2, 3]. Affected by Dabashan uplift from west to east, the coal forming period of coal measures of Xujiahe Formation was gradually postponed from west to east[4]. The coal-bearing horizon from Guangyuan to Kaijiang gradually increases from west to east, so that the coal forming environment shifts from west to east. In Guangwang Coalfield, the lower coal measures of Xujiahe Formation are mainly mined, while the upper coal measures are mainly mined in Dazhou and Kaijiang.

2. 2. 2 Huayingshan Coalfield in the East of Sichuan Province

Huayingshan Coalfield is mainly mined by Guangneng Company and the Upper Permian Longtan Formation belongs to the main minable coal-bearing strata. Around the Late Permian coal swamp in Huayingshan Mountain, highlands of the ancient land around Dabashan, Leshan-Longnvsi uplift, and ancient highlands of Hannan are found[5-7]. These contain areas of terrigenous clasts in the coal-accumulating basin in the Late Permian Huayingshan Mountain. The coal-bearing strata mainly show the tidal-flat sedimentary environment of a bay (lagoon) in sedimentary areas of marine and continental transitional facies and a sedimentary environment of carbonate platforms in shallow-sea sedimentary areas[8], including a set of sandstone and argillaceous rock intercalated with thin layers of limestone. Underlying the coal-bearing strata are Maokou limestone from volcanic ash in the middle Permian series[9]. The overlying strata are the Upper Permian Changxing Formation (a set of limestone rocks sedimented in a carbonate platform). The strata are in conformable contact[10, 11].

The Longtan Formation with an average thickness of 136m contains either one or two coal seams with an average total thickness of 2. 40m. Coal seam K_1 at the bottom of the Longtan Formation is mainly mined and shows a thickness of 0. 8m to 3. 57m in the Huayingshan Coalfield. Coal seam K_1 is formed in an intertidal environment[12]. The coal seam is forked into K_{1-1} and K_{1-2} layers. Grey to grey-white mafic tuff with a thickness of about 2m to 5m, is found at the bottom of coal seam K_1 and shows conchoidal fractures and seems slippery. It is in unconformable contact with the limestone in

Maokou Formation (Early Permian) underlying the bottom of the Longtan Formation. By studying symbiotic (associated) elements in coal in the Huayingshan mining area[13], it is considered that Huayingshan coal seam has the characteristics of less volatile bituminous coal, in which the amounts of kaolinite, quartz, pyrite, and calcite are relatively high.

2.2.3 Coalfield in the South of Sichuan Province

The coalfield in the south of Sichuan Province is mined by Furong Company and its Xuanwei Formation and Late Permian Longtan Formation belong to the most mined coal-bearing strata. This area has complex and diverse strata and the Upper Permian Formation is divided into the Wujiaping Stage and Changxing Stage. The strata in the Wujiaping Stage include Emeishan Basalt Formation and Longtan Formation (the middle and lower members of the Xuanwei Formation or Wujiaping Formation), while those in the Changxing Stage consist of the Changxing Formation strata (upper member of the Xuanwei Formation). The Longtan Formation and the middle and lower members of the Xuanwei Formation or the Wujiaping Formation represents heterogeneous deposition at a similar time, and the Changxing Formation and the upper member of the Xuanwei Formation also present similar deposition as above. The three formations in the Upper Permian are an Emeishan Basalt Formation, Longtan Formation (the middle and lower members of the Xuanwei Formation or Wujiaping Formation), and the Changxing Formation (the upper member of the Xuanwei Formation) from bottom to top[14-16].

Furong Company mainly mines coal seams in Furong mining area and Guxu mining area. In Furong mining area, coal-bearing strata are of the Xuanwei Formation with marine and continental alternative facies in the Late Permian strata, with an average thickness of 130m. The strata in the Xuanwei Formation comprise siltstone, sandstone, mudstone, sandy mudstone, argillaceous limestone, clay rock, and coal seams. The coal-bearing strata are divided into the upper, middle, and lower coal-bearing members. The main coal seam lies in coal measure B in the middle member, including coal seams B_2, B_3, and B_4 from bottom to top[17] and coal seam $C_5(B_4^{upper})$ in the upper member.

In the Guxu mining area, coal-bearing strata are the alternate marine-continental strata dominated by continental facies in the Upper Permian Longtan Formation, with an average thickness of 100m. The Longtan Formation mainly includes mudstone, claystone, sandy mudstone, siltstone, argillaceous siltstone, and fine sandstone. It is intercalated with coal seams and carbonaceous mudstones. There are between 6 and 15 coal seams, of which four are mineable and mostly mineable. The top is intercalated with thin layers of bioclastic marlite, containing many plant fossils and a few animal fossils. At the bottom, there are scattered, nodular and dendritic pyrites, which are enriched into layers.

2.2.4 Panzhihua Coalfield

The Panzhihua Coalfield is mainly mined by Panzhihua Coal Company and coal-bearing strata are mainly mined in the Triassic Daqiaodi Formation. The strata lie in the middle member of the Kangdian ancient land at the western margin of the Yangtze Paraplatform and distributed in the rhombic block in Sichuan, Yunnan, and Guizhou Provinces and contain many fractures[18]. After intense magmatic activities from the Late Permian to Early Triassic, a series of nearly N-S trending graben basins were formed in the whole Kangdian axis under the background of continuous E-W trending extensional structures. The east side of the Panzhihua fracture rises to a horst and the west side descends to a graben, forming Baoding and Hongni graben basins in the mining area[19]. After uplifting and denudation in the Middle Triassic, the original Baoding and Hongni basins sank into basins again in the early stage of the Late Triassic, and a set of coal-bearing molasse formations was deposited therein, namely the Daqiaodi Formation[18].

The Daqiaodi Formation mainly consists of conglomerate, siltstone, coarse to medium to fine sandstone, carbonaceous shale, and shale intercalated with limestone and coal seams. Cyclothems are developed and more than 100 cyclothems with unequal thicknesses (coarse at the bottom and fine at the top) are found, with obvious top and bottom boundaries, so they are easily identified in the field[20]. The lower part of the Daqiaodi Formation is sandstone and mudstone intercalated with coal seams with swamp facies of rivers and lakes; The middle part is interbedded with sandstone and conglomerate alternated by braided rivers and alluvial fans. Moreover, the upper part presents ultra-thick conglomerates with alluvial fan facies. In addition, there are 96 coal seams in this formation, with a total thickness of 52.61m, in which 49 coal seams are minable or partially minable, with a total thickness of 31.25m.

2.3 Distribution of Geological Conditions

The different coal-forming processes and complicated geological structures result in diverse occurrence conditions of coal seams. Under such condition, it is necessary to reflect the occurrence conditions of coal seams in this area and to provide a reliable guarantee of geological resources for technological and theoretical research into GER application. For this purpose, the geological factors, such as thickness and dip angle of coal seams, buried depth of roadways, thickness and lithology of immediate roofs, and roof integrity in 143 mining roadways in four coalfields are statistically analysed over a period of four months. The parameters are ranked, in descending order, as shown in Figures 2.1~2.6.

2.3.1 Thicknesses of Coal Seams

The distribution of thicknesses of coal seams is shown in Figure 2.1.

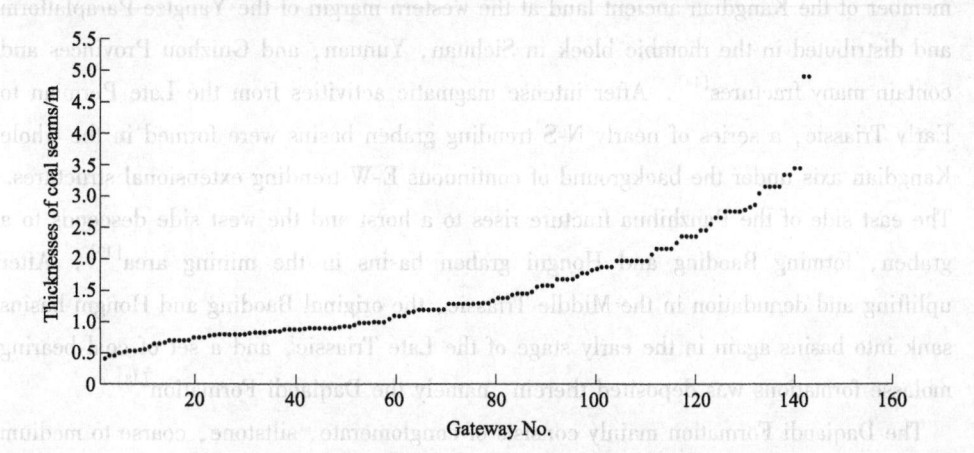

Figure 2.1 Distribution of thicknesses of coal seams

2.3.2 Dip Angles of Coal Seams

The distribution of dip angles of coal seams is shown in Figure 2.2.

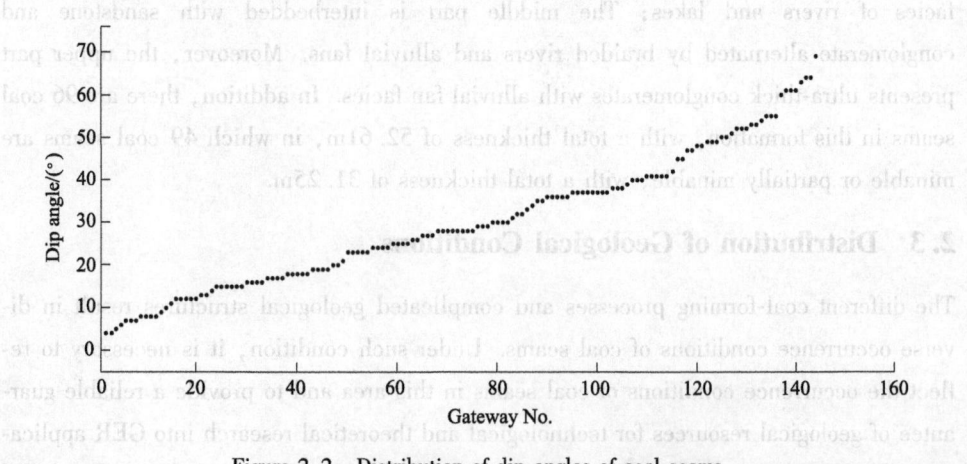

Figure 2.2 Distribution of dip angles of coal seams

2. 3. 3 Burial Depths of Roadways

The distribution of buried depths of the roadway is shown in Figure 2. 3.

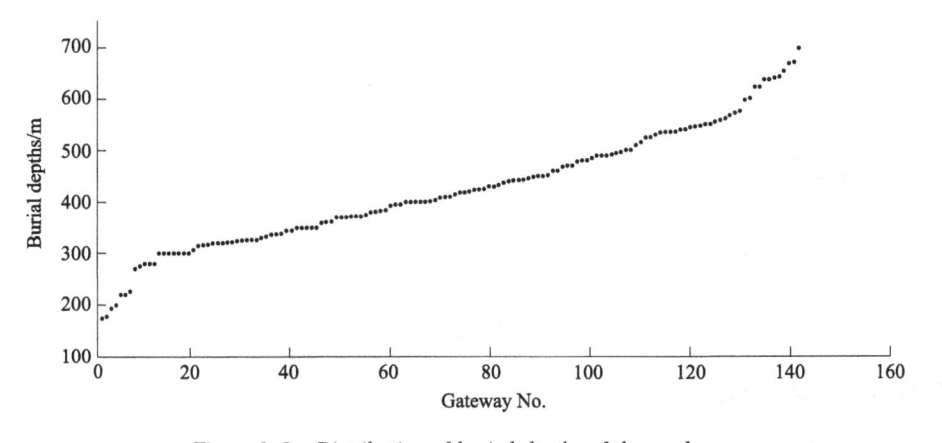

Figure 2. 3 Distribution of buried depths of the roadways

2. 3. 4 Thicknesses of Immediate Roofs

The distribution of thicknesses of immediate roofs is shown in Figure 2. 4.

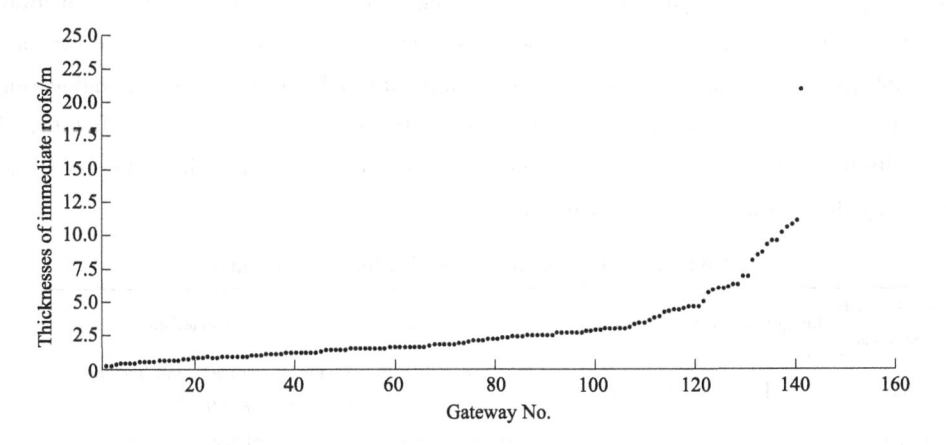

Figure 2. 4 Distribution of thicknesses of immediate roofs

2. 3. 5 Lithologies of Immediate Roofs

Lithology generally reflects the rock strength, so the Protodyakonov coefficient f is mainly considered when discussing the lithology of immediate roofs, as it characterises the strength of roof strata. The distribution of lithologies (f values) of immediate roofs is shown in Figure 2. 5.

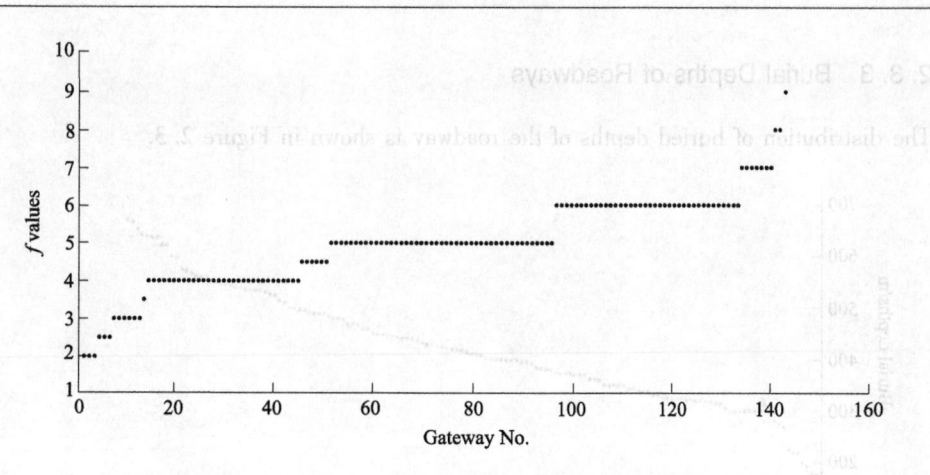

Figure 2.5　Distribution of lithologies (f values) of immediate roofs

2.3.6　Roof Integrity

The rock mass itself has high compressive and tensile strength, but due to the presence of structural planes, such as joints and fractures, the integrity of rock mass deteriorates, and the strength thereof decreases, so that it is easily damaged under the influence of mining. Practice has proven that it is difficult to support broken roofs with poor integrity and support costs are high. Therefore, roof integrity has an important effect on roadway stability. At present, elastic wave velocity (integrity coefficient K_v) is used to evaluate roof integrity, but it is expensive, time-consuming and laborious to apply in engineering practice. To reduce cost, improve engineering efficiency, and ensure the reliability of integrity data, expert scoring method was used to determine the integrity of the rock surrounding the roadways, as shown in Table 2.2[21].

Table 2.2　The integrity classification of rock mass

Classification of rock mass	Integrity level	Integrity score	Engineering geological characteristics
Intact	I	(80,100]	The spacing between joints (beddings) and fractures is larger than 1.5m (K_v>0.75)
Relatively intact	II	(70,80]	The spacing between joints (beddings) and fractures is smaller than 1.5m, but larger than 1m ($0.55<K_v \leqslant 0.75$)
Medium	III	(60,70]	The spacing between joints (beddings) and fractures is smaller than 1m, but larger than 0.4m ($0.35<K_v \leqslant 0.55$)
Broken	IV	(50,60]	The spacing between joints (beddings) and fractures is smaller than 0.4m, but larger than 0.1m ($0.15<K_v \leqslant 0.35$)
Very broken	V	≤50	The spacing between joints (beddings) and fractures is smaller than 0.1m ($K_v \leqslant 0.15$)

2.3 Distribution of Geological Conditions · 19 ·

The geological and technical personnel scored the roof integrity of the roadways as follows Figure 2.6.

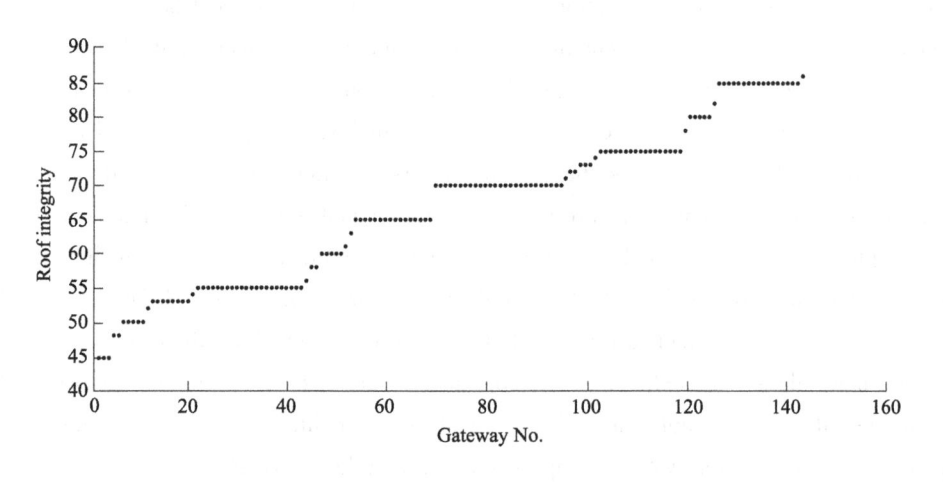

Figure 2.6 Distribution of roof integrity

As shown in Figures 2.1~2.6:

(1) Thicknesses of coal seams. According to statistical data, there are only two coal seams with a maximum thickness of 5m in the mining roadways; The rest have a maximum thickness smaller than or equal to 3.5m, which accounts for 98.6% of the seams, that is, the majority of coal seams are thin and medium-thick coal seams.

(2) Dip angles of coal seams. The dip angles of coal seams vary widely. The minimum dip angle is smaller than 5°, while the maximum is close to 70°. Near-horizontal, gently inclined, inclined and steeply inclined coal seams are distributed throughout the area of interest.

(3) Burial depths of roadways. The burial depths of the roadways are less than 700m, most of which are concentrated between 300m and 600m, basically being mined in the shallow part[22].

(4) Thicknesses of immediate roofs. The minimum and maximum thicknesses of immediate roofs are 0.2m and 21m respectively, and most of them are smaller than 2.5m.

(5) Lithology of immediate roofs. Most values of the Protodyakonov coefficient f are greater than or equal to 4, reflecting the fact that the uniaxial compressive strengths of rock in immediate roofs are generally high.

(6) Roof integrity. The roofs of a few mining roadways contain strata in Grade I. The integrity of roofs in other mining roadways is below 80 (with most being about 70, or below), representing rock masses of medium grade and below, showing poor integrity.

2.4 Chapter Summary

We have analysed the distribution and formation process of coal-bearing strata in Sichuan Coal Group. Due to complex paleogeographic environment and changes, and geological structures in the late stage, there are rich geological resources of coal seams in the four coalfields. Most of coal seams in Sichuan Coal Group are thin and medium-thick, with a thickness of less than 3.5m. Near-horizontal, gently inclined, inclined and steeply inclined coal seams are distributed throughout the area of interest. The mining depth ranges from less than 200m to 700m; The thicknesses of immediate roofs change continuously, and most of immediate roofs are thinner than 2.5m. The rock strengths of immediate roofs are generally high, but only a few immediate roofs comprise a rock mass classified as Grade I. The rock masses are of medium grade or below, with poor integrity. The complex and diverse geological conditions provide a guarantee of geological resources for systematic application and studies of GER.

References

[1] Dai C C. Study on sequence filling patterns and reservoir distribution of the xujiahe formation in sichuan foreland basin[D]. Chengdu: Chengdu University of Technology, 2011.

[2] Li Y J, LiangW L, Shao L Y. Late triassic coal-bearing strata sequence stratigraphy and coal accumulation characteristics in sichuan province[J]. Coal Geol China, 2011, 23(8): 32-37.

[3] Li Y J. Sequence-palaeogeography and coal accumulation of the late triassic xujiahe formation in the sichuan basin[D]. Beijing: China University of Mining and Technology, 2014.

[4] Huang QS. Paleoclimate and coal-forming characteristics of the late triassic xujiahe stage in northern sichuan[J]. Geol Rev, 1995, 41(1): 92-99.

[5] Tang Y G, Ren D Y, Liu Q F, et al. Relationship between the coal forming environment and sulfur in the late permian coal of Sichuan, China[J]. Acta Sedimentologica Sinica, 1996, 14(4): 162-168.

[6] Xie J R, Li G H, Luo F Z. Reservoir characteristics of the upper triassic xujiahe formation in Sichuan Basin, China[J]. J Chengdu U Technol, 2009, 36(1): 13-18.

[7] Shao L Y, Gao C X, Zhang C, et al. Sequence-palaeogeography and coal aaccumulation of late permian in southwestern China[J]. Acta Sedimentologica Sinica, 2013, 31(5): 856-866.

[8] LuoY B. The study of trace elements geochemistry in Late Permian coal and tuff samples from east and south Sichuan province, China[D]. Beijing: China University of Mining and Technology, 2014.

[9] Dai S, Wang X, Zhou Y,et al. Chemical and mineralogical compositions of silicic, mafic, and alkali tonsteins in the late Permian coals from the Songzao Coalfield, Chongqing, Southwest China [J]. Chem Geol, 2011, 282(1,2): 29-44.

References

[10] Liu G M, Li D L. Groundwater movement in the changxing limestone aquifer at the longtan coalfield, huaying mountain[J]. Acta Geol Sichuan, 2009, 29(4): 470-473.

[11] Zhang Z R, Zhang Z M, Li P, et al. Analysis on Geological Structure Features and Gas Deposit Law in Lüshuidong Mine[J]. Coal Sci Technol, 2009, 37(10): 105-108,128.

[12] Zhang T, Tang H, Wu B. Sedimentary facies of the carboniferous system at the north pitching end of the huayingshan anticline[J]. Acta Geol Sichuan, 2010, 30(3): 271-274.

[13] Zhuang X, Su S, Xiao M, et al. Mineralogy and geochemistry of the Late Permian coals in the Huayingshan coal-bearing area, Sichuan Province, China[J]. Int J Coal Geol, 2012, 94: 271-282.

[14] Gao C X. Sequence-palaeogeography and Coal-accumulation of Late Permian in Chuan-Yu-Dian-Qian China[D]. Beijing: China University of Mining and Technology, 2015.

[15] Gong S L, Zhang C X. Sequence stratigraphic features and coal——accumulating laws of permian basin in Southern China[J]. Coal Geol China, 2001, 13(2): 10-12,117.

[16] Huang N H, Wen X D, Wang G F. Sedimentary structures of tidal flat in the permian coal——bearing sequences of eastern yunnan and western quizhou[J]. Earth Sci, 1985, 10(4): 63-70.

[17] Jiang X T, Liu X. Geologic feature of coal and gas outburst in baijiao coal mine[J]. J Jiaozuo I Technol, 1997, 16(6): 18-19,21-25.

[18] Chen Z M. Depositional environment of the late triassic daqiaodi formation, panzhihua[D]. Chengdu: Chengdu University of Technology, 2014.

[19] Lu J, Shao L Y, Wei K M, et al. Paleogeographical evolution and coal accumulation in a sequence stratigraphic framework in the Baoding fault basin of western Yangtze paraplatform[J]. J China Coal Soc, 2009, 34(4): 433-437.

[20] Shao L Y, Lu J, Ran, L M, et al. Late Triassic sequence stratigraphy and coal accumulation in Baoding Basin of Sichuan Province[J]. J Palaeogeog, 2008, (4): 355-361.

[21] Yuan L, Xue J H, Liu Q S. Surrounding rock stability control theory and support technique in deep rock roadway for coal mine[J]. J China Coal Soc, 2011, 36(4): 535-543.

[22] Liang Z G. The question about the boundary division of deep of shallow part's coal mining[J]. J Liaoning Tech U, 2001, 20(4): 554-556.

3 Adaptation Assessment of Gob-Side Entry Retaining

3.1 Introduction

Since 1950s, pillarless gateways have been widely used in underground coal mining industry, mainly in the UK, Germany, Poland, Russia and China, and extensive studies have been carried out with regards to different geological conditions. It is known that many factors impact the quality of gob-side entry retaining, including geological conditions and detailed retaining technique. Geological factors include the cover depth, mining height, coal seam dip angle, etc., while the detailed retaining technique involves the support method, the cross section size and shape, etc. In mining, the former is objective and beyond the control of mining practitioners and the latter is subjective and determined by the mining practitioners. Further study has shown that the detailed retaining technique is basically controlled by natural factors. For example, the gateway shape mainly depends on the size and dip angle of coal seam, and the support method is dependent on lithology of surrounding rock mass. Therefore, the study on geological factors is significant in gob-side entry retaining.

Usually, the geological conditions of coal seam are complex, such as in China. So far, with different geological conditions, a large number of coal mines have used the technique of gob-side entry retaining, such as cover depths[1-3], different mining heights[4,5], different roof lithologies[6-8] and different coal seam dip angles[9,10]. Such cases have provided good experience for future gob-side entry retaining design. However, many impact factors are often described as uncertain variables and there is a lack of system that can weigh the relative importance of these impact factors. What is more, the adaptation of gob-side entry retaining is fuzzy and a grade system has not been developed. This leads to the inconvenience when utilizing this roadway technique in practice.

In rock mechanics and rock engineering, some rock mass classification systems have been developed and successfully used in tunnelling, underground mining based on practical experience, such as Rock Quality Designation (RQD), Rock Mass Rating scheme (RMR), Rock Mass Quality Index (Q-system), China Technical Standard for

Bolt Shotcrete Support (GBJ 86-85) and coal mine deep rock roadway support design classification[11]. These help the engineer to qualitatively understand the potential behaviour of surrounding rock mass during engineering and further provide guidance for support design. Recently, classification of the surrounding rock mass for the purpose of gateway support design has been widely studied[12,13].

To promote the application of gob-side entry retaining in underground coal seam mining and reduce the support cost due to inappropriate decision, this study presents a statistical analysis of retained gateways in China mining industry and distribution characteristics of geological factors are summarized. Then, Analytic Hierarchy Process (AHP) was applied to weighting these factors[14]. Another step is to invite experienced field practitioners, i. e. , senior engineers in China, to mark with the 1-9 scale method[15]. Subsequently, the adaptive grades and scope of each grade are investigated and the support method for each grade is provided according to relative regulations, statistical data and practical experience. Finally, a decision-making method is proposed to determine the adaptation grade for a gateway and take the appropriate support method based on fuzzy comprehensive evaluation.

3.2 Determination of Impact Factors

3.2.1 Determination Principles

The factors that influence the gob-side entry retaining are various due to site-specific geological conditions. To make it representative, some principles should be followed when selecting these factors:

(1) Materiality principle. The selected factors should have a significant effect on the stability of gateway, excluding the secondary factors that have relatively small effect on stability.

(2) Independence principle. One factor can reflect an aspect of the respective attributes, and the correlation among other factors should be as low as possible.

(3) Separability principle. Obvious difference exists in sample data among different factors[16].

(4) Easy acquisition principle. Factors can be easily measured or acquired in the coal mine and quantitative data can be provided for the design.

(5) Fundamental principle. The selected factors should be fundamental to facilitate comparative analysis for practitioners using monitoring data in the process of construction or after construction, even from the support system unfavourably selected.

(6) Universal principle. Selected factors should be universal in mining area.

3. 2. 2 Factors Selection

Based on the determination principles mentioned above and combining mining geological conditions, mining technology, and relevant practical experience, six geological factors were selected to evaluate the adaptation of gob-side entry retaining: coal seam dip angle (α), mining height (m), cover depth (H), thickness influence coefficient of immediate roof (N), lithology of the immediate roof and roof integrity, as shown in Figures 3. 1 and 3. 2.

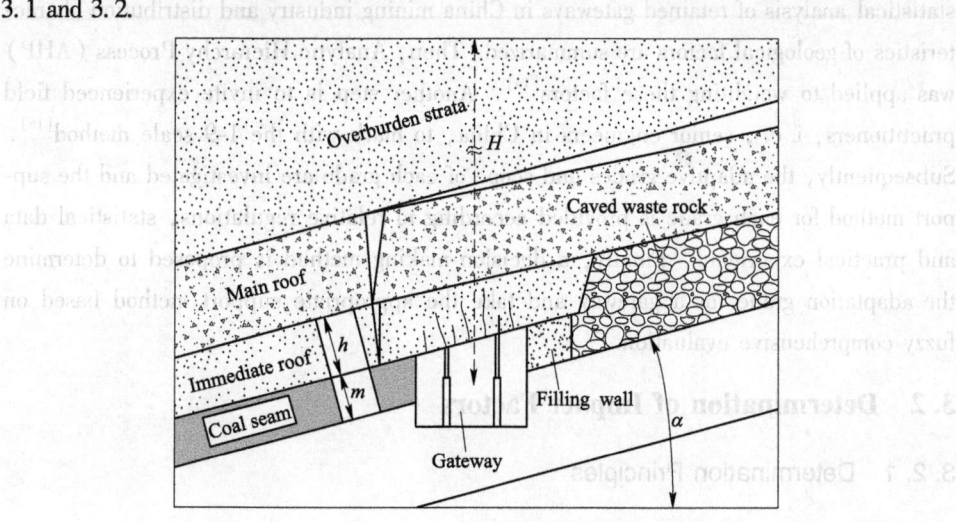

Figure 3. 1 Schematic diagram of gob-side entry retaining

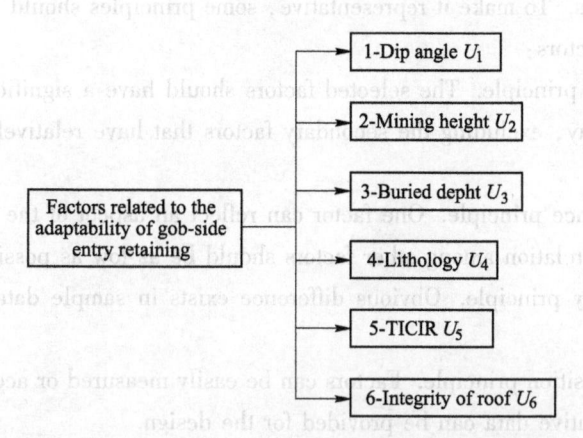

Figure 3. 2 Factors selected and used in assessing the adaptability of gob-side retaining technique

For the dip angle of coal seam (α), it mainly influences the stress distribution in the surrounding rock. With the change of coal seam dip angle, the stress in two side walls

3.2 Determination of Impact Factors

and roof of the gateway changes greatly, leading to the difference in failure mode and size of failure zone of surrounding rock. Moreover, if the dip angle is greater than the natural repose angle of gangue in the gob, specific measures should be taken to prevent gangue fleeing.

Here, mining height (m) denotes the thickness of coal seam. Generally speaking, the cave-in scope of immediate roof and main roof increases with the coal seam thickness. This could lead to the difficulty in maintaining the retained gateway as the height increasing in collapsed roof strata intensifies the degree of periodic weighting. Moreover, the height of the constructed filling wall should be based on the thickness of the coal seam and higher filling wall is needed for larger mining height. This not only requires extra filling materials but also increases the difficulty for filling process and reduce the stability of the developed filling wall.

The in-situ stress generally increases with the cover depth, indicating more energy stored in surrounding rock mass before mining. The mining excavation leads to energy release from surrounding rock mass and tends to transfer to the unmined coal block. Therefore, the cover depth increase can make the failure zone larger and the retained gateway difficult to maintain.

The lithology of immediate roof determines its strength. The higher the strength of the immediate roof, the greater the roof-cut resistance needed on the gob side. This requires complex technology to cut the hard immediate roof, otherwise the deadweight of roof overhang in the gob area would transfer the roof of the retained gate, leading the roof damage. On the other hand, if the immediate roof has very lower strength, it can be broken easily and also hard to support.

The thickness influence coefficient of the immediate roof (N) mainly impacts the stability of main roof after the immediate roof cave-in. It can represent the stability degree of the immediate roof to the retained gateway for a certain mining height (m) and can be expressed as follows:

$$N = \frac{h}{m} \tag{3.1}$$

Existing study shows that the caved immediate roof can fill the gob and support overburden strata timely when $N>4$, which can reduce the bearing pressure of the filling wall hand hence is good for retained gateway stability; When $N<4$, the caved block of immediate roof is insufficient to fill the gob, leading to main roof bends, cracks and rotates to subside until touching the caved waste rock. As a result, the filling wall has to experience dynamic loading of overlying rock strata movement, leading to the retained gateway hard to maintain. In such cases, special support scheme is needed, however, the induced supporting cost is high.

3   Adaptation Assessment of Gob-Side Entry Retaining

Regarding the roof integrity (T), the structural surfaces, such as a joint and fissure, dominate in mechanical behavior of roof strata. Weak roof integrity represents the case difficult to support or high support cost. Moreover, if the propagation of fracture reaches the aquifer can lead to water inflow, even inundation. For rock mass of strong water-weakening properties, the inflow water can significantly reduce the rock strength and eventually damage the self-support capacity of roof strata.

In the following sections, the factors mentioned above are treated as the evaluation indicators in weight analysis and fuzzy comprehensive evaluation.

3.3   Field Application Statistics

Gob-side entry retaining has been widely used in mining industry, China. To obtain the current and detailed utilization information of this technique in terms of six proposed geological factors, 145 retained gateways were investigated and the distribution characteristics of each factors were analyzed, as shown in Figures 3.3~3.7. It is noted that not all geological factors were obtained in each of investigated retained gateways due to site-specific conditions.

3.3.1   Dip Angle

Figure 3.3 shows the distribution of coal seam dip angle over 126 retained gateways. A four-segment line is used to represent the change in the number of retained gateways

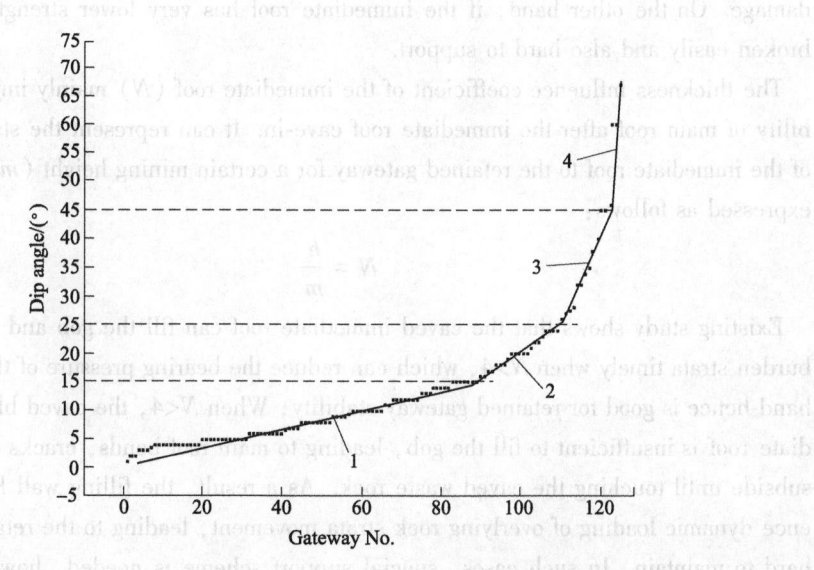

Figure 3.3   Change of dip angle in retained gateways

of different dip angle, where larger slope denotes greater change of the dip angle. It can be seen that three transit points (15°, 25° and 45°) can be found, corresponding to the tendency of the line segments to change. Figure 3.3 shows that most of the dip angles were under 25°, accounting for 88.1%. By statistics, few gateways with large dip angle in Figure 3.3 indicates that it is hard for gateway retaining for coal seam of high dip angle, illustrating that the dip angle of coal seam has a great influence on the adaptation of gob-side retaining technique.

3.3.2 Mining Height

Figure 3.4 summarizes 145 retained gateways under different mining height condition. Based on the 3 line segments as shown in Figure 3.4, two transit points, 2.5m and 3.5m, can be found. Also, thickness of coal seam mostly used the adaptation of gob-side entry retaining technique are under 2.5m, accounting for 77.2%, few over 3.5m. This indicates that the retaining of headgate may be difficult for thick coal seams.

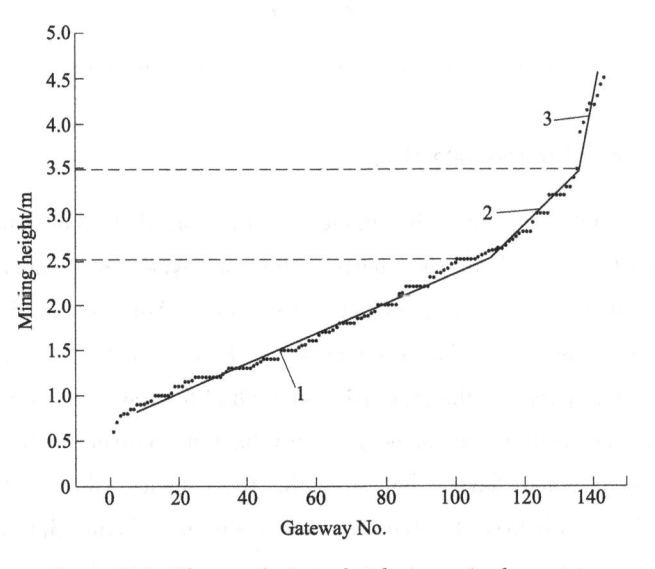

Figure 3.4   Change of mining height in retained gateways

3.3.3 Cover Depth

There are only 80 retained gateways reflecting the cover depth information in retained gateways, as shown in Figure 3.5, indicating that cover depth may be not an important factor influencing the adaptation of gob-side entry retaining technique. It can be seen that the cover depth changes from 90m to nearly 1000m, and there are two transit points, 300m and 700m, and most retained gateways are conducted at the depth of 300~700m.

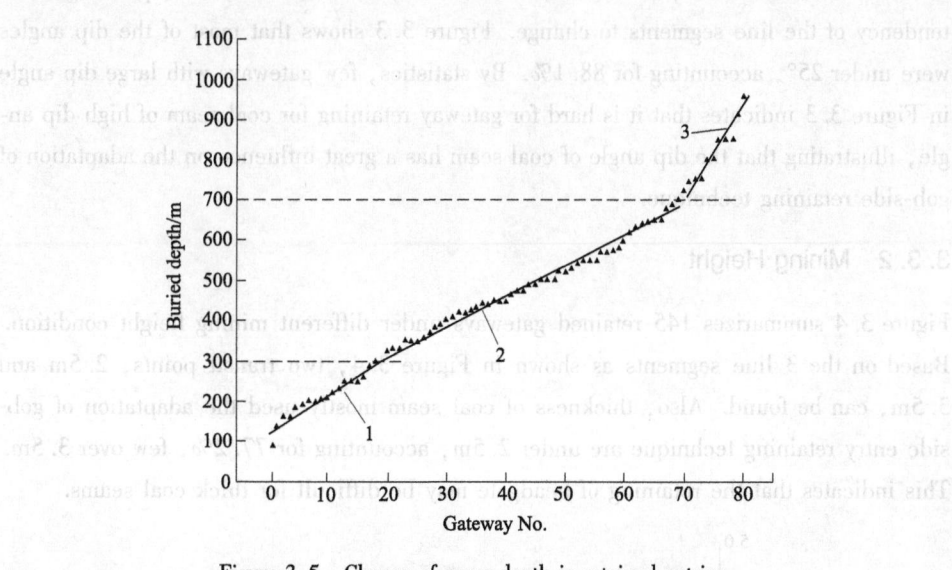

Figure 3.5   Change of cover depth in retained entries

3.3.4   Lithology of Immediate Roof

According to the statistical data, the lithology of the immediate roof mainly consists of nine types of rock. To facilitate the analysis, the rock types are numbered as follows: 1-limestone, 2-siltstone, 3-sandy mudstone, 4-gritstone, 5-mudstone, 6-medium sandstone, 7-fine sandstone, 8-argillaceous sandstone, 9-sandy shale. It is well known that the lithology generally reflects the rock mass strength of the roof. For example, the limestone, gritstone and medium sandstone generally have a relatively high strength, while mudstone and sandy shale have a low strength. By statistics, the number of retained gateways of different lithology of immediate roof is shown in Figure 3.6. Obviously, the successful retained gateways were not only carried out under hard immediate roof conditions, but also goes well under soft immediate roof conditions, reflecting that the lithology may has little influence on this retaining technique.

3.3.5   TICIR

Figure 3.7 shows that the distribution of TICIR in 124 retained gateways. It can be seen that most of the influence coefficients were under 4, illustrating that the caved immediate roof cannot fill the gob and that the cracking and rotation of main roof occur in the

3. 3   Field Application Statistics   · 29 ·

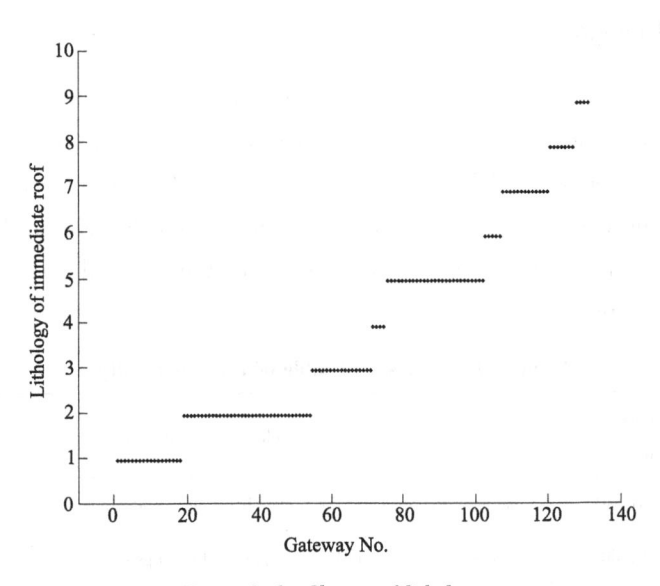

Figure 3. 6   Change of lithology

process of periodic roof weighting. However, many successful cases regarding gob-side
entry retaining with TICIR larger than 4 indicate that the current technology can overcome the impact of main roof movement and hence TICIR does not play a dominant role
in gob-side entry retaining.

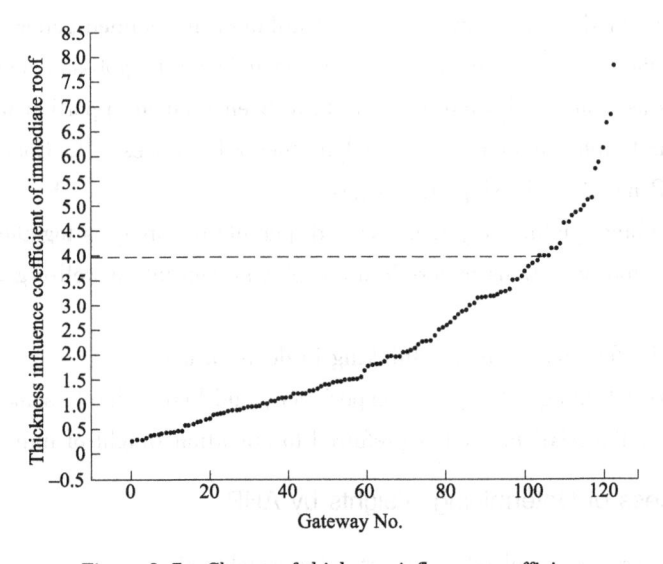

Figure 3. 7   Change of thickness influence coefficient

3.3.6 Roof Integrity

Regarding the influence of roof integrity on adaptation of gob-side entry retaining, there is no quantitative indicator that can reflect the roof integrity for the gateways concerned, leading to no data being available for in-depth analysis. However, the integrity of rock mass can be reflected by joint spacing and thus can be used to index roof integrity, as shown in Table 3.1[11]. As it can be conveniently measured in field, it is expected to be widely used for gateway retaining.

Table 3.1 Assessment table of roof integrality

Integrity grade	Mark of integrity T	Geologic characteristic
I	[100, 80]	Spacing $l \geqslant 1.5$m between joints (beddings) and cracks
II	(70, 80]	Spacing 1m$\leqslant l < 1.5$m between joints (beddings) and cracks
III	(60, 70]	Spacing 0.4m$\leqslant l < 1$m between joints (beddings) and cracks
IV	(50, 60]	Spacing 0.1m$\leqslant l < 0.4$m between joints (beddings) and cracks
V	$\leqslant 50$	Spacing $l < 0.1$m between joints (beddings) and cracks

3.4 Indicator Weighting

The results presented above show the factor distribution in retained gateways, but cannot quantitatively show the relative importance of each factor in gob-side entry retaining. Many methods associated with multi-factors have been used to apportion weight of indicators, such as Delphi Method[17] and analytic hierarchy process[18]. Compared to other methods, AHP has the following advantages:

(1) it can combine qualitative judgments and quantitative analysis together.

(2) it can decompose various factors in a complex system into a well-organized network diagram.

(3) it meets the features of human thinking in decision-making.

(4) it can take advantage of experts' experience, and knowledge to some extent[19,20].

In this study, the AHP method is preferred to apportion weight of indicators.

3.4.1 Process of Determining Weights by AHP

The following steps are involved in apportioning weights of indicators:

(1) Obtaining mark results. The 1-9 scale method is used to obtain mark results[20].

3.4 Indicator Weighting

By comparison, relative importance among different factors are determined by experienced field practitioners. The marking criteria are shown in Table 3.2 and the factors are shown in Figure 3.2.

Table 3.2 Marking criteria

Mark of i compared to j	Equally important	Weakly importance	Strongly important	Very strongly important	Extremely important
	1	3	5	7	9

Comment: 2, 4, 6, 8 are the intermediate values between two adjacency values; If mark i to j is p_{ij}, then mark j to i is $p_{ji} = 1/p_{ij}$.

(2) The pairwise judgment matrix (P) is obtained according to the mark results:

$$P = (p_{ij})_{n \times n} = \begin{pmatrix} p_{11} & p_{12} & \cdots & p_{1n} \\ p_{21} & p_{22} & \cdots & p_{2n} \\ \vdots & \vdots & \vdots & \vdots \\ p_{n1} & p_{n2} & \cdots & p_{nn} \end{pmatrix} \tag{3.2}$$

Where p_{ij} is the importance degree of the ith factor compared to the jth factor defined which indicates the ratio scale of the score from the experts; n represents the dimension of the judgment matrix.

(3) The following formula is used to normalize the elements of matrix P:

$$p'_{ij} = \frac{p_{ij}}{\sum\limits_{k=1}^{n} p_{kj}}, \quad i,j = 1,2,\cdots,n \tag{3.3}$$

Then, the normalization matrix, P', can be acquired with:

$$P' = (p'_{ij})_{n \times n}$$

(4) Conducting summation of the elements of the same line/row of normalization matrix P' and expressed as:

$$w'_i = \sum\limits_{j=1}^{n} p'_{ij}, \quad i = 1,2,3,\cdots,n \tag{3.4}$$

(5) The weight vector $W = (w_1, w_2, \cdots, w_n)$ is then developed by:

$$w_i = \frac{w'_i}{\sum\limits_{k=1}^{n} w'_i}, \quad i = 1,2,3,\cdots,n \tag{3.5}$$

(6) The maximum eigenvalue λ_{max} is computed as follows:

$$\lambda_{max} = \frac{1}{n} \sum\limits_{i=1}^{n} \frac{(P'W^T)_i}{w_i}, \quad i = 1,2,\cdots,n \tag{3.6}$$

(7) Finally, a consistency check is applied by computing the consistency index (CI):

$$CI = \frac{\lambda_{max} - n}{n - 1} \tag{3.7}$$

Where CI is the consistency index, where small value of CI represents great consistency. The CI becomes zero if the matrix completely meets the requirements. Considering that the deviation of consistency may be caused by random reasons, the consistency ratio can be computed as:

$$CR = \frac{CI}{RI} \tag{3.8}$$

Where RI is the random index, which changes with variations in the dimensions, as shown in Table 3.3 When $CR \leqslant 0.10$, it means that the consistency of the pairwise judgment matrix is acceptable in engineering.

Table 3.3 RI Values

Dimension n	1	2	3	4	5	6	7	8	9
RI Values	0	0	0.58	0.9	1.12	1.24	1.32	1.41	1.45

3.4.2 Results of Weights

The more mark results there are, the more reliable the weights are. Thus, eighty experts who had rich practice experience were invited to mark. One of these pair-wise comparisons is shown here as an example:

$$P = \begin{pmatrix} 1 & 2 & 3 & 2 & 3 & \dfrac{1}{2} \\ \dfrac{1}{2} & 1 & 3 & 1 & 3 & \dfrac{1}{3} \\ \dfrac{1}{3} & \dfrac{1}{3} & 1 & \dfrac{1}{2} & \dfrac{1}{3} & \dfrac{1}{4} \\ \dfrac{1}{2} & 1 & 2 & 1 & 2 & 2 \\ \dfrac{1}{3} & \dfrac{1}{3} & 3 & \dfrac{1}{2} & 1 & \dfrac{1}{3} \\ 2 & 3 & 4 & \dfrac{1}{2} & 3 & 1 \end{pmatrix} \tag{3.9}$$

Then, taking the mode that is highest frequently appeared from the mark results, a comprehensive normalization pair-wise comparison matrix was built:

$$P' = \begin{array}{c} \begin{array}{cccccc} U_1 & U_2 & U_3 & U_4 & U_5 & U_6 \end{array} \\ \begin{array}{c} U_1 \\ U_2 \\ U_3 \\ U_4 \\ U_5 \\ U_6 \end{array} \begin{pmatrix} 0.273 & 0.388 & 0.176 & 0.300 & 0.225 & 0.218 \\ 0.136 & 0.193 & 0.176 & 0.300 & 0.225 & 0.218 \\ 0.090 & 0.064 & 0.059 & 0.050 & 0.025 & 0.055 \\ 0.136 & 0.097 & 0.176 & 0.150 & 0.225 & 0.218 \\ 0.090 & 0.064 & 0.176 & 0.050 & 0.075 & 0.073 \\ 0.273 & 0.194 & 0.235 & 0.150 & 0.225 & 0.218 \end{pmatrix} \end{array} \tag{3.10}$$

Subsequently, the maximum eigenvalue was calculated, i. e. , $\lambda_{max} = 6.257$, and the value of CR was calculated by:

$$CR = 0.041 < 0.1 \tag{3.11}$$

Therefore, it can be judged that there is a good consistency for the matrix. The factor weights are summarized in Table 3.4.

Table 3.4　The weights of geological factors of gob-side entry retaining

Factors	Dip angle	Mining height	Buried depth	Lithology	TICIR	Integrity of roof
Weights	0.263	0.208	0.056	0.167	0.088	0.216

It can be observed that the weight of the dip angle is highest and becomes the most important geological factor of adaptability of gob-side entry retaining technique, the roof integrity and mining height are ranked as the second and third, respectively, while the cover depth is of the lowest weight and becomes the least important factor. It can also be found that the weight difference between roof integrity and mining height is small.

3.5　Adaptive Grades and Their Support Methods

In this section, the adaptive grade and the support method of each grade are analyzed to facilitate decision-making in gob-side entry retaining.

3.5.1　The Analysis of Support Methods and Their Utilization Conditions

The safety of a gob-side retained gateway depends on basic support, reinforced support and gob-side wall[21-23]. After reviewing previous retained gateways, the support methods were analysed statistically. Basic support ensures the integrity of surrounding rock mass such that there are less fractures developed in the early stage; the main supports are bolts and cables with steel mesh. Meanwhile, reinforced support decreases the deformation of a roof in the abutment pressure zone and mining face during mining. Except high strength and high pre-stressed bolts and cables, individual hydraulic props with articulated roof beams are used as a part of main support[24].

Gob-side wall can withhold the deformation during roof weighting to some extent and hence can help maintain entry stability during mining process. Seven ways have been used to construct the gob-side wall: high-water material (HWM), concrete, paste, block[23], waste pack, caved waste rock (CWR) and advanced roof pre-splitting and cutting (ARPC)[25,26]. It is worth noting that no gob-side wall is needed when using the technique of advanced roof pre-splitting and cutting or the gob-side wall is constructed just to isolate the gateway from gob area without the capacity to bearing load.

Even though the coal seam dip angle, roof integrity and mining height have relative high weight value in gob-side entry retaining. However, it is found that the roof integrity has no specific information contributing to the support method on the gob side and cannot be deeply analysed. Therefore, only the support methods with the factors of dip angle and mining height were dissected.

3.5.1.1 The Support Method for Different Coal Seam Dip Angle

The method is analysed as shown in Figure 3.8.

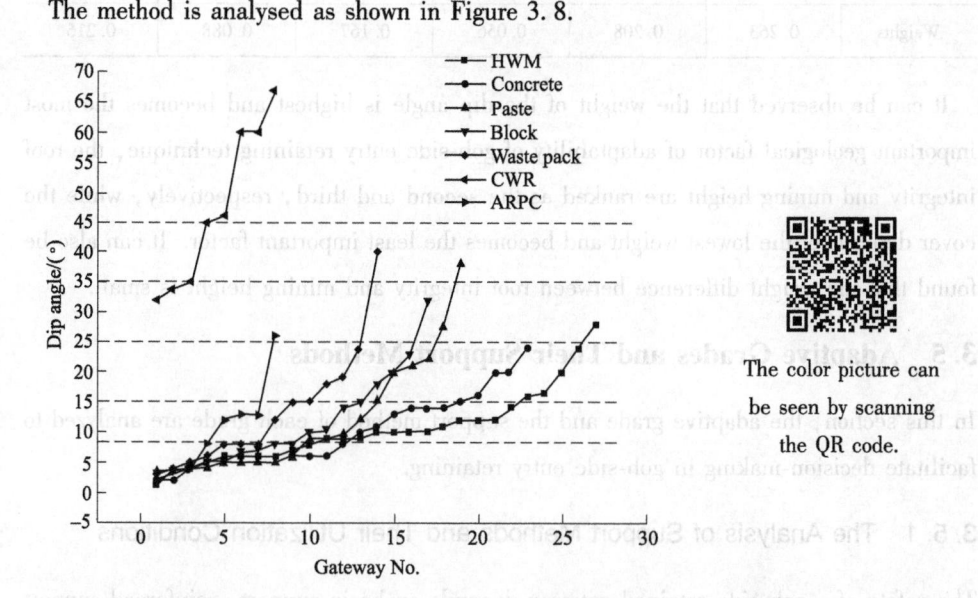

Figure 3.8   The rules of the support method with dip angle

Combining the standard coal seam dip angles (flat seam: less than 8°, gently inclined seam: less than 25°, inclined seam: 25°~45°, steeply pitching seam: more than 45°) with Figures 3.3~3.8, the angle α can be divided into 6 intervals: $\alpha \leqslant 8°$, $8°<\alpha \leqslant 15°$,

$15°<\alpha\leqslant25°$, $25°<\alpha\leqslant35°$, $35°<\alpha\leqslant45°$ and $\alpha>45°$. When $\alpha>45°$, only caved waste rock was used; When $35°<\alpha\leqslant45°$, the support can be one of CWR, paste and HWM; When $25°<\alpha\leqslant35°$, there were almost 6 ways to construct gob-side wall, but the mature methods are paste, block and waste pack; When $15°<\alpha\leqslant25°$, there were five mature support methods, except CWR and ARPC; And when $\alpha\leqslant15°$, there were almost six support methods, except CWR.

The results illustrate that the support methods and the number of retained gateways greatly changed with the coal seam dip angle. A larger number of gateway is retained at coal seam of small dip angle and also the diverse support methods are diverse at small coal seam dip angle. The above data also indicate that when the dip angle was less than the waste rock natural repose angle (almost $35°\sim40°$), the gateway can be retained relatively easily for small dip angle cases. Moreover, the CWR method's service dip angle is more than $30°$, and becomes more and more concerned and acceptable[27] because of the good physical properties of high compressive strength and great compressibility[28-30] of the caved waste rock, making it as an ideal filling material. Until now, CWR waste rock was used in the coal seam in longwall mining and flexible shield mining in the false dip, which usually had a small coal pillar on the gob-side.

3.5.1.2 The Support Method for Different Mining Height

The method are analysed and shown in Figure 3.9.

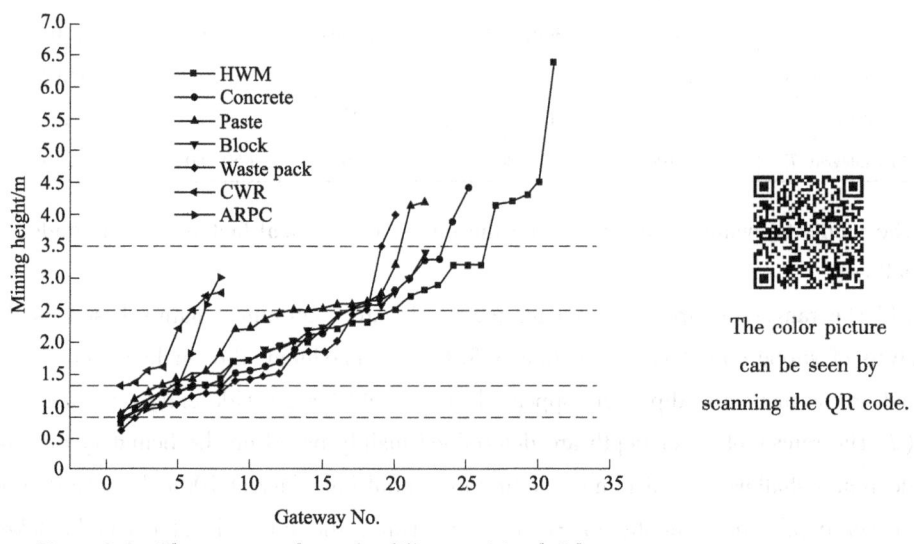

The color picture can be seen by scanning the QR code.

Figure 3.9 The support scheme for different mining height

When the thickness of coal seam is less than 0.4m, the mining is infeasible economically. Combining the standard thickness of coal seams (extra thin seams: less than 0.8m, thin seam: less than 1.3m, medium-thickness coal seam: 1.3 ~ 3.5m, thick seam: more than 3.5m), Figures 3.4 ~ 3.9, the mining heightcan be divided into five intervals regarding the adaptability of gob-side entry retaining: $0.4m < m \leqslant 0.8m$, $0.8m < m \leqslant 1.3m$, $1.3m < m \leqslant 2.5m$, $2.5m < m \leqslant 3.5m$ and $m > 3.5m$. When $m > 3.5m$, there were four support methods except block, CWR and ARPC; When $2.5m < m \leqslant 3.5m$, there were five mature methods except CWR and ARPC; When $1.3m < m \leqslant 2.5m$, all the methods can be used; When $0.8m < m \leqslant 1.3m$, there are almost seven methods; And when $0.4m < m \leqslant 0.8m$, there are four mature methods: ARPC, HWM, concrete and waste pack.

3.5.2 Adaptive Grades

Here, these six geological factors are used to grade the adaptive degree of gob-side retaining. Five adaptive grades were chosen. The ranges of adaptive factor for each grade are given in Table 3.5.

Table 3.5 Adaptive grades and their factor ranges

Grades	I (Very easy)	II (Easy)	III (Medium)	IV (Hard)	V (Very hard)
Dip angle α	$\leqslant 8°$ or $(65°, 75°]$	$(8°, 15°]$ or $(55°, 65°]$	$(15°, 25°]$ or $(45°, 55°]$	$(35°, 45°]$	$(25°, 35°]$
Mining height m	$(0.4, 0.8]$	$(0.8, 1.3]$	$(1.3, 2.5]$	$(2.5, 3.5]$	>3.5
Cover depth H	$\leqslant 300$	$(300, 500]$	$(500, 700]$	$(700, 900]$	>900
Lithology (f value)	>6	$(3, 6]$	$(1.5, 3]$	$(0.5, 1.5]$	$\leqslant 0.5$
TICIR N	>4	$(3, 4]$	$(2, 3]$	$(1, 2]$	$\leqslant 1$
Roof integrity T	$(80, 100]$	$(70, 80]$	$(60, 70]$	$(50, 60]$	$\leqslant 50$

The points to evaluate the grades and the range of geological factors of each grade are as follows:

(1) The ranges of dip angle and mining height are determined mainly based on the analysis of support methods in section 3.5.1. The maximum dip angle is set as 75° based on the maximum dip angle appeared in the field for gob-side entry retaining.

(2) The ranges of cover depth are determined mainly based on the boundary division of deep and shallow in coal mines, setting the initial boundary at 700m[31] according to the cover depth characteristic in Figure 3.5. Thus, the interval value can be taken as 200m.

(3) Regarding the lithology of immediate roof, the Protodyakonov coefficient (f) is employed to express the strength, which is defined as $f=\sigma_c/10$, where σ_c is the uniaxial compressive strength of roof rock. One may think that the immediate roof may be too hard to be cut off on the gob side and the gateway may be difficult to retain if f is high. However, the method of retaining gateways can be updated with the technology of advanced roof pre-splitting and cutting (ARPC)[25,26]. The range of lithology (f value) and roof integrity T are determined based on the Standard for Classification of Engineering Rock Masses (GB 50218—94) in China[11].

(4) The ranges of TICIR, N, are determined mainly based on the practical experience provided by field practitioners.

3.5.3 Support Method of Each Grade

According to the analysis of support methods and adaptive grades, a set of supporting schemes is recommended, which is given in Table 3.6. The basic support and reinforced support in gateways can be used for each grade, while a specific support method can be chosen based on economic consideration.

Table 3.6 **Support schemes for each grade of gob-side entry retaining**

Adaptive grades	Support method on gob-side	Support method in gateways
I	HWM, concrete, waste pack, ARPC	Basic support:
II	HWM, concrete, waste pack, ARPC, block, paste, CWR	High strength and high pre-stress bolt and cable with mesh Reinforced support:
III	HWM, concrete, wastepack, CWR, block, paste	(1) Individual hydraulic prop with articulated roof beam; (2) Specially designed strengthened hy-
IV	Paste, block, waste pack	draulic support; (3) High strength and high pre-stressed
V	HWM, paste	bolt and cable

3.6 Adaptability Evaluation Process

Fuzzy mathematics has the capability to reflect the real world[32] and it is a powerful mathematical tool to modelling uncertain systems in industry, nature and humanity, as well as facilitating common-sense reasoning in decision making in the absence of complete and precise information[18].

To study the adaptive grade of gob-side entry retaining for site-specific conditions and

· 38 ·　　　3　Adaptation Assessment of Gob-Side Entry Retaining

further provide a support decision-making for a gateway of specific geological conditions, fuzzy comprehensive evaluation theory is used here for further analysis. The first step is to establish membership function, i. e., calculating the membership degree of each factor to each grade. Then, evaluation matrix is calculated, where a single factor determines a fuzzy mapping relationship vector, and multiple factors constitute the fuzzy relational matrix, which is the evaluation matrix expressed by:

$$R = \left\{ \begin{matrix} r_{11} & \cdots & r_{1m} \\ \vdots & \ddots & \vdots \\ r_{n1} & \cdots & r_{nm} \end{matrix} \right\} \qquad (3.12)$$

Subsequently, compositional operation is conducted. After the compositional operation of the weight set (W) and evaluation matrix, the comprehensive evaluation result vector (B) can be obtained as follows:

$$\begin{cases} B = WR = (b_1, b_2, b_3, \cdots, b_n) \\ R = (r_{ij})_{m \times n}, \quad r_{ij} \in [0, 1] \end{cases} \qquad (3.13)$$

Furthermore, the values in vector B are calculated by using the weighted average method:

$$b_j = \sum_{i=1}^{n} (W_i r_{ij}), \quad j = 1, 2, \cdots, m \qquad (3.14)$$

Lastly, the grade is determined. Find the maximum value $b_j (j = 1, 2, \cdots, n)$ in vector B as the evaluation result, and determine the corresponding grade of the evaluation object according to the maximum attached principle.

3.6.1　Determination of Membership Function

The values of the evaluation indicator and the determination of membership in comprehensive evaluation are key and difficult to obtain. Due to the complexity and diversity of six geological factors, the selection of evaluation factor and determination of membership must be close to the actual situation as much as possible. Five types of distribution function are adopted based on the random interval ideology.

The numerical characteristics analysis method provides the idea that an indicator exists in each category interval, and if the value of the quantitative factors just equal to the indicator, the probability of the factor belonging to the category is treated as 1. Other values of the factors maybe distribute between the indicators of each category, and the closer to the indicator, the greater the possibility of the factor belonging to the category[33]. Thus, the membership function obviously has normal distribution and the modified normal distribution function is used as the membership function. The indicator values are the mean of normal distribution and given as follows:

$$\mu_j(x_i) = \exp\left[-\left(\frac{x_i - m_{ij}}{\sigma_{ij}}\right)^2\right], \quad j = 1,2,\cdots,s \tag{3.15}$$

Where s represents the numbers of categories. The normal distribution curves are shown in Figure 3.10.

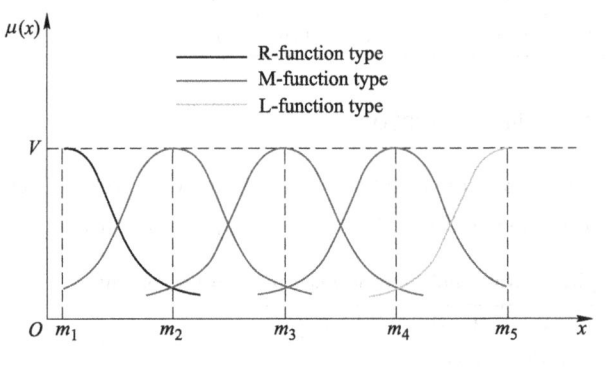

The color picture can be seen by scanning the QR code

Figure 3.10 The normal distribution curves

Where $\mu_j(x_i)$ is the membership degree of the element i to j; x_i is the value of the ith element among the evaluation elements; m_j and σ_j are the statistical constants. And m_j is the cluster center value for each category, which means that the probability of the factor belonging to the category is 1; the value of σ_j can be determined as follows:

$$\mu_j(x_i) = \exp\left[-\left(\frac{x_u - x_d}{\sigma_j}\right)^2\right] = 0.5 \tag{3.16}$$

Where x_u, x_d——the boundary value of each grade range.

Equation (15) illustrates that when the indicator equals to the boundary value, the membership degree is 0.5 to the two adjacent grades, and σ_j can be calculated as:

$$\sigma_j = \frac{x_d}{0.833}, \quad \text{R-function type} \tag{3.17}$$

$$\sigma_j = \frac{|x_u - x_d|}{1.665}, \quad \text{M-function type} \tag{3.18}$$

$$\sigma_j = \frac{x_u}{0.833}, \quad \text{L-function type} \tag{3.19}$$

Therefore, the values of m_j and σ_j of each grade can be obtained, as listed in Table 3.7.

· 40 ·   3 Adaptation Assessment of Gob-Side Entry Retaining

Table 3. 7 Coefficient m and σ membership functions

Factors	U_1		U_2		U_3		U_4		U_5		U_6	
	m_{1j}	σ_{1j}	m_{2j}	σ_{2j}	m_{3j}	σ_{3j}	m_{4j}	σ_{4j}	m_{5j}	σ_{5j}	m_{6j}	σ_{6j}
I	0/70	9. 6/6	0. 6	0. 24	200	120	7	1. 8	5	1. 2	100	12
II	11. 5/60	4. 2/6	1. 1	0. 3	400	120	4. 5	1. 8	3. 5	0. 6	75	6
III	20/50	6	1. 9	0. 72	600	120	2. 25	0. 9	2. 5	0. 6	65	6
IV	40	6	3	0. 6	800	120	1	0. 6	1. 5	0. 6	55	6
V	30	6	4	0. 6	1000	120	0	0. 6	0	1. 2	45	6

3. 6. 2 Actual Gateway Adaptability Evaluation

To validate the evaluation method, six representative retained gateways were selected. Six relevant geological factors are listed in Table 3. 8 assessed by geological engineers.

Table 3. 8 Six geological factors and their support method on gob side

Factors	Dip angle	Mining height	Buried depth	Lithology	TICIR	Support method on gob side
Case 1 (Jinggang mine)	11	2. 1	489	4	0. 92	Concrete
Case 2 (Xuyong mine)	16	0. 9	300	2	1. 1	Block (waste pack)
Case 3 (Xiaobaoding mine)	20	2. 2	450	5	1. 45	Block
Case 4 (Huashan mine)	38	2. 6	460	6	1. 15	CWR
Case 5 (Bolin mine)	25	0. 48	520	3	2. 81	ARPC
Case 6 (Taiping mine)	60	2. 78	200	5	3. 23	CWR

Through calculation, the comprehensive evaluation vectors B_1, B_2, B_3, B_4, B_5 and B_6 of the six gateways are as follows:

$$
\begin{cases}
B_1 = [\,0.019 \quad 0.523 \quad 0.346 \quad 0.058 \quad 0.052\,] \\
B_2 = [\,0.072 \quad 0.306 \quad 0.309 \quad 0.240 \quad 0.070\,] \\
B_3 = [\,0.040 \quad 0.200 \quad 0.420 \quad 0.296 \quad 0.042\,] \\
B_4 = [\,0.100 \quad 0.217 \quad 0.200 \quad 0.407 \quad 0.074\,] \\
B_5 = [\,0.370 \quad 0.180 \quad 0.316 \quad 0.001 \quad 0.100\,] \\
B_6 = [\,0.165 \quad 0.488 \quad 0.178 \quad 0.163 \quad 0.003\,]
\end{cases}
\qquad (3.20)
$$

According to the principle of maximum membership, the adaptive grades of six cases are II, III, III, IV, I and II respectively, and the support method on gob-side matches the corresponding method of adaptive grades very well. Therefore, this method can be taken as a scientific way in assisting the retaining design of gob-side.

In addition to the six retained gateways mentioned above, six geological factors of another 149 gateways in coal mine, Sichuan Province, China, were collected during this study, refer to attachment. Using the adaptive grades of the gateways proposed in this study, the supporting schemes perform well in field and are well accepted by field practitioner.

3.7 Chapter Summary

In this study, six geological factors to evaluate the adaptation of gob-side entry retaining are discussed and their weights were calculated. Finally, the adaptation grades of gob-side entry retaining technique under site-specific geological conditions and their support methods are also proposed. In detail, following the principles of materiality, independence, separability, easy acquisition, foundational and universality, six geological factors were proposed to evaluate the adaptability of gob-side retaining technique including coal seam dip angle, mining height, cover depth, lithology of immediate roof, TICIR and roof integrity. Using the 1-9 scale method of AHP and 80 experts' mark results, indicator weight of six geological factors were obtained, where the coal seam dip angle was identified as the most important factor, while cover depth was marked as the least important. Finally, five adaptive grades were proposed, and their support methods are also recommended using the fuzzy comprehensive evaluation theory. Moreover, the framework to determining the adaptive grade for a specific gateway was provided. The validity of the proposed method was confirmed using six field cases. Therefore, the proposed method and framework can be considered as a supplementary tool in retaining design of gob-side entry in addition to traditionally used rock mass classification methods.

References

[1] Tang J X, Deng Y H, Tu X D, et al. Analysis of roof separation in gob-side entry retaining combined support with bolting wire mesh[J]. J China Coal Soc, 2010, 35: 1827-1831.

[2] Chen Y, Bai J B, Wang X Y, et al. Support technology research and application inside roadway of gob-side entry retaining[J]. J China Coal Soc, 2012, 37: 903-910.

[3] Zhang N, Chen H, Chen Y. An engineering case of gob-side entry retaining in one kilometer-depth soft rock roadway with high ground pressure[J]. J China Coal Soc, 2015, 40: 494-501.

[4] Zheng X G, Zhang N, Yuan L, et al. Method and application of simultaneous pillar-less coal mining and gas extraction by staged gob-side entry retaining[J]. J China U Min Techn, 2012, 41: 390-396.

[5] Xue J H, Han C L. Strata behavior and control countermeasures for the gob-side entry retaining in

the condition of large mining height[J]. Min Saf Eng, 2012, 29: 466-473.

[6] Cao S G, Zou D J, Bai Y J, et al. Study on upward mining of sublevels for gob-side entry retaining in three-soft thin coal seam group[J]. Min Saf Eng, 2012, 29: 322-327.

[7] Ning J G, Ma P F, Liu X S, et al. Supporting mechanism of "yielding-supporting" beside roadway maintained along the goaf under hard rocks[J]. Min Saf Eng, 2013, 30: 369-374.

[8] Cheng Y H, Jiang F X, Lin J K, et al. Experimental study on gob-side entry retaining by roadside flexible packing under hard roof[J]. Min Saf Eng, 2012, 29: 757-761.

[9] Hua X Z, Ma J F, Xu T J. Study on controlling mechanism of surrounding rocks of gob-side entry with combination of roadside reinforced cable supporting and roadway bolt supporting, and its application[J]. Rock Mech Eng, 2005 24: 2107-2112.

[10] Zang C, Zhang L. Analysis of roof cutting working resistance of gob-side entry retaining in steep coal seam[J]. Coal Techn, 2015, 34: 68-70.

[11] Liu Q, Gao W, Yuan L. Stability control theory and supporting technology of deep rock roadway in coal mine and its application[M]. Beijing: Science Press, 2010.

[12] Chen D, Hua X, Li Y, et al. Study of key technologies and management modes for classifying surrounding rocks of gateway[J]. Rock Mech Eng, 2012, 31: 2240-2247.

[13] Zhang N, Wang C, Gao M, et al. Roadway support difficulty classification and controlling techniques for Huainan deep coal mining[J]. Rock Mech Eng, 2009, 28: 2421-2428.

[14] Saaty T L. The analytic hierarchy process[D]. New York: McGraw Hill, 1980.

[15] Wang Y, Yang W, Li M, Liu X. Risk assessment of floor water inrush in coal mines based on secondary fuzzy comprehensive evaluation[J]. Rock Mech. Min. Sci,2012, 52: 50-55.

[16] Zhang L, Pan J N, Zhang X M. Fuzzy comprehensive evaluation of mining geological condition in the No.9 coal seam, Linhuan coal mine, Huaibei Coalfield, China. Procedia[J]. Environ Sci, 2012,12 A: 9-16.

[17] Hoseinie S H, Ataei M, Osanloo M. A new classification system for evaluating rock penetrability [J]. Rock Mech Min Sci, 2009, 46: 1329-1340.

[18] Ataei M, Mikaeil R, Hoseinie S H, et al. Fuzzy analytical hierarchy process approach for ranking the sawability of carbonate rock[J]. Rock Mech Min Sci, 2012, 50: 83-93.

[19] Han K, Li F, Li H, et al. Fuzzy Comprehensive Evaluation for Stability of Strata over Gob Influenced by Construction Loads. Energ. Procedia[J]. 2012, 16 B: 1102-1110.

[20] Tang Y, Sun H, Yao Q, et al. The selection of key technologies by the silicon photovoltaic industry based on the Delphi method and AHP (analytic hierarchy process): Case study of China [J]. Energy, 2014, 75: 474-482.

[21] Hua X Z. Study on gob-side entry retaining technique with roadside packing in longwall top-coal caving technology[J]. Coal Sc Eng, 2004, 10: 9-12.

[22] Kang H, Niu D, Zhang Z, et al. Deformation characteristics of surrounding rock and supporting technology of gob-side entry retaining in deep coal mine[J]. Rock Mech Eng, 2010, 29: 1977-1987.

[23] Tan Y L, Yu F H, Ning J G, et al. Design and construction of entry retaining wall along a gob-

side under hard roof stratum[J]. Rock Mech Min Sci, 2015, 77: 115-121.

[24] Yuan L. Gas distribution of the mined-out side and extraction technology of first mined key seam relief-mining in gassy multi-seams of low permeability [J]. China Coal Soc, 2008, 33: 1362-1367.

[25] Zhang G F, He M C, Yu X P, et al. Research on the technique of no-pillar mining with gob-side entry formed by advanced roof caving in the protective seam in Baijiao coal mine[J]. Min Saf Eng, 2011, 28: 511-516.

[26] Xue J H, Duan C R. Technologies of gob-side entry retaining with no-pillar in condition of overlying and thick-hard roof[J]. China Coal Soc, 2014, 39: 378-383.

[27] Zhou B, Xu J, Zhao M, et al. Stability study on naturally filling body in gob-side entry retaining [J]. Min Sci Techn, 2012, 22: 423-427

[28] Jiang Z Q, Ji L J, Zuo R S. Research on mechanism of crushing-compression of coal waste[J]. China U Min Techn, 2001, 30: 139-142.

[29] Canibano J. Construction of an experimental embankment using coal mining wastes[C]. The 4th International Symposium on Reclamation, Treatment and Utilization of Coal Mining Wastes, Cracow, 1993: 335-342.

[30] Liu S Y, Qiu Y, Tong L Y, et al. Experimental study on strength properties of coal wastes[J]. Rock Mech Eng, 2006, 25:199-205.

[31] Liang Z G. The question about the boundary division of deep of shallow part's coal mining[J]. Liaoning Techno U, 2001, 20: 554-556.

[32] Ertuğrul İ, Tuş A. Interactive fuzzy linear programming and an application sample at a textile firm[J]. Fuzzy Optim Decis Ma, 2007, 6: 29-49.

[33] Li S. Engineering Fuzzy Mathematics and Its Application[M]. Harbin: Harbin Institute of Technology Press, 2004.

4 Advancing Cutting Roof for Retaining Gateway With Near Horizontal and Close Coal Seam Groups

4.1 Introduction

In recent years, many scholars have carried out considerable research on mechanical models and the structural design of gob-side supporting structures, and they have obtained many conclusions. The main gob-side artificial filling walls currently in use include wooden stacks, densely spaced hydraulic props, gangue walls, concrete walls, high-water packing material, and other fill materials[1,2]. The construction of these above-mentioned structures on the gob-side in a roadway is mainly completed by people. These structures belong to the category of additional artificial supporting structures. Nevertheless, building an artificial filling wall requires a more complicated computation of its bearing capacity and spending more time preparing the support material and doing construction. Sometimes, the material for an artificial filling wall is very expensive.

As technology had advanced, advanced cutting roofs for gob-side entry retaining were invented, which needed no artificial filling wall[3,4]. In addition, this technology changed passive 'support' to active 'cut' and 'support' to roof rock mass, making the ground pressure for profit. The technology was first successfully applied to the headgate of a 2422 working face with a hard limestone roof in the Baijiao coalmine in Sichuan province, China[5]. The technology then had a generalized application in China after the year 2010. The Sunzhuang coalmine headgate of the 12465 Ⅱ working face was retained using this technology to cut a hard limestone roof. In addition, the Nantun coalmine gateway with a hard limestone roof for the 1610 working face was cut and retained[6]. Beyond that, the Tangshangou coalmine tailgate of an 8820 working face with a thick-bedded sandstone of roof was effectively cut along the gob, forming a stabilized gateway[7]. The Bolin coalmine headgate of a 0456(K24) working face was retained through this method, and the surrounding rock deformation decreased significantly compared to the method of constructing an artificial filling wall[8]. A coalmine with a coal seam of 1.6m and a hard immediate roof of 3.78m in thickness was chosen to study the engineering application of the cutting roof approach. The roof was still complete and flat, the gangue rib was integrally formed with the help of the gob-side support, and the entry was retained with high quality after mining[3]. For the Suncun coalmine with a 31120 working face and a 2415 working face, the coal seam had an average thickness of 3.0m, the immediate roof was siltstone with an average thickness of 5.3m, and the cutting roof for the retaining gateway achieved good field application results. For the Xiashanmao coalmine, the 8102 and 9101 working faces with a thickness of 12.9m and an limestone immediate roof with a thickness of 3.78m were successfully retained by roof cutting without a pillar[2]. The Xinchao coalmine with a 90101 panel had a thick and hard roof retained by roof cutting, successfully ensuring the panel safe production[9]. The Jining No. 3 coalmine with a 5312 panel based on the geological conditions of hard roof carried out roof blasting and cutting, and verified the pressure-relief effect[10]. It can be seen that the technology of the retaining gateway was usually applied in the gateways with hard or harder roof rock masses, or in coal seams that had a single geological condition[2,7,11-15]. Although the technology for retaining gateways is frequently used in the field, the related experimental research in laboratories has not been carried out. This has resulted in many structures with inner roof rock mass being misunderstood.

Considering the coal seam dip angle is the most influential geological factor on the adaptability of retaining gateway. Experimental methods were used to research a roof structure based on the geological conditions of flat seam and close distance coal seam group in the Baijiao coalmine, China. Thus, the aim of this chapter was to fully comprehend the roof stability mechanism of an advancing cutting roof for gob-side entry retaining, with consideration of equivalent material simulation experiments that had been widely used in the deformation and failure research of the strata and gateways of surrounding rock[16-18]. Therefore, Section 4.2 of the paper describes the mechanism of a cutting roof for retaining gateways. Section 4.3 of the paper describes the geological and mining information and the modeling procedures of the equivalent material model available for the case study. The obtained simulation results are covered in Section 4.4 of the paper. Sections 4.5 and 4.6 end the paper with discussion and conclusions.

4.2 Mechanism of the Cutting Roof for a Retaining Gateway

An advancing cutting roof for gob-side entry retaining involves several procedures. First, drilling and blasting is performed for the roof at a certain distance in the original rock stress zone ahead of the working face, and a cutting roof plane is formed whose direction is consistent with the working face advancing direction. A cutting roof line the distance of which to the high side wall was usually 50mm, is shown in Figure 4.1[5]. As the working face ad-

· 46 · 4 Advancing Cutting Roof for Retaining Gateway With Near Horizontal and Close Coal Seam Groups

vances, the front abutment pressure and the periodic weighting will cut the roof along the cutting roof plane. The cut roof will form a caving roof wall behind the working face. This will eliminate the influence on the hanging roof to the gateway surrounding rock, and the retained gateway roof will form a cantilever beam structure.

Simultaneously, the gob caving waste rock will support the overburdened strata, and control the main roof as well as the rotation and weighting deformation. By reducing the abutment pressure at the same time, this will also reduce the dynamic disaster possibilities of rock burst and coal and gas outburst in a coalmine, as well as other disaster possibilities. Furthermore, the support in the retained gateway will preserve the integrity of the roof rock mass and prevent roof separation[5-7,19].

Furthermore, in a cutting roof design, the two parameters cut height (h) and cut line deviation angle (γ), as shown in Figure 4.1(c), play an important role in the roof behavior and the effect of the retained gateway[6].

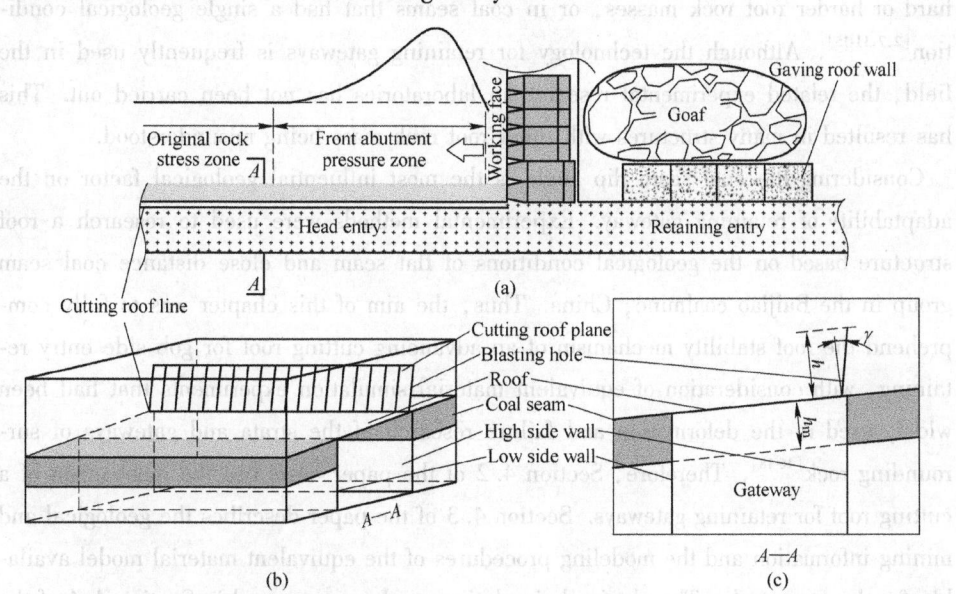

Figure 4. 1   Model of an advancing cut roof for gob-side entry retaining
(a) Plane graph; (b) parts of the model structure; (c) cross-section diagram.

4. 3   Equivalent Material Simulation Experiment

In order to gain an in-depth understanding of advancing cutting roof for gob-side entry retaining, including the stabilizing mechanism and caving structure, we carried out a plane stress experiment on an equivalent material model based on the geological conditions of a flat seam and a close distance coal seam group in the Baijiao coalmine, Sichuan Province, China.

4.3.1 Geological Conditions

Permianperiod strata is the coal-bearing strata of Baijiao coalmine, and theprimary working coal seams are #4, #3, #2, and the top layering of the #2 coal seam, as shown in Figure 4.2. After a detailed analysis and discussion with engineers, the #24mining district

Order No.	Thickness /m	Lithology
1		Shale
2	0.6	Limestone
3	2.0	Top layering of #2 coal seam
4	0.8	Clay rock
5	8.8	Sandy mudstone
6	0.2	Coal streak
7	2.3	Sandy mudstone
8	1	Clay rock
9	1.8	Sandy mudstone
10	2	Fine sandstone
11	2	Sandy mudstone
12	0.4	Coal streak
13	1.7	Fine sandstone
14	1.75	Sandy mudstone
15	1.1	#2 coal seam
16	4.5	Clay rock
17	1.3	#3 coal seam
18	0.4	Clay rock
19	2.1	Fine sandstone
20	1.9	#4 coal seam
21	1.2	Clay rock
22	1.0	Sandy mudstone
23	1.8	Fine sandstone
24	1.2	Sandy mudstone
25	0.6	#5 coal seam
26	2.4	Clay rock
27		Fine sandstone

(a)   (b)

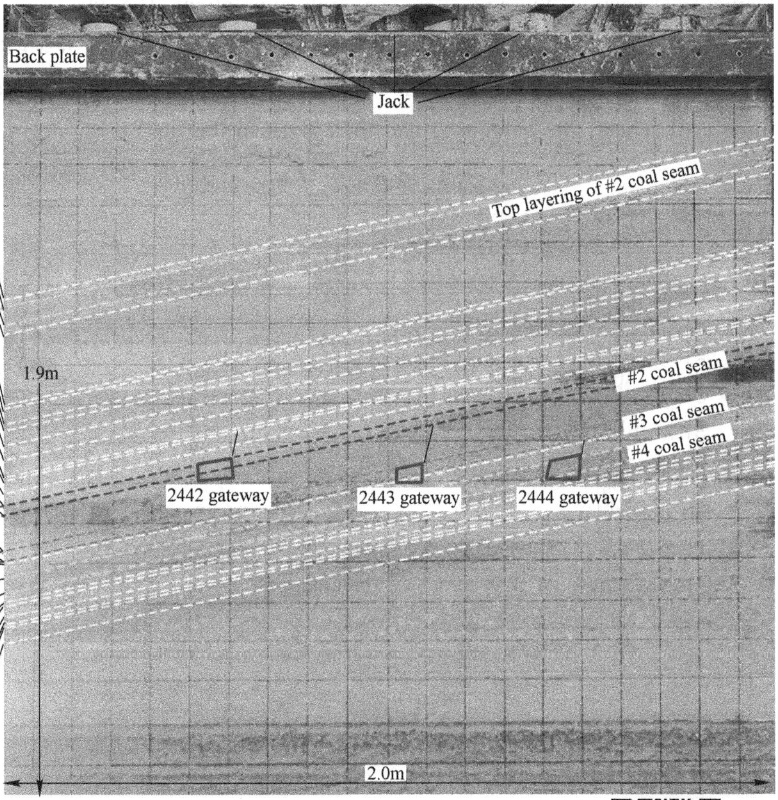

Figure 4.2 Schematic of the coal-bearing strata

(a) Formation lithology; (b) Equivalent material model

The color picture can be seen by scanning the QR code.

was chosen as the geological background because the three gateways in that mining district had been retained though the cutting of the roofs. Therefore, we conducted a survey of the geological conditions and found that the average thicknesses of the four coal seams were 1.9m, 1.3m, 1.1m and 2.0m. The average distances between the adjacent coal seams were 2.5m, 4.5m, and 22.75m from the bottom up, and the average coal seam dip angle was 10°. The coal seams belonging to the flat seam and the close distance coal seam group, as well as the lithology of the coal-bearing strata in the study area, are shown in Figure 4.2(a).

Among the four coal seams, #2, #3, and #4 were identified as the coal and gas outburst coal seams. The top layering of the #2 coal seam was treated as the protective coal seam and it was mined in the mining district to produce an unloading pressure effect below the three coal seams. Eventually, the #2, #3, and #4 coal seams were chosen as the experimental coal seam of the equivalent materials simulation. The 2442 headgate, 2443 headgate, and 2444 headgate in each coal seam were the preparatory test gateways for cutting the roof and retaining, as shown in Figure 4.2(b). The relationship between the below three coal seams, the gateways, and their sizes in the field is shown in Figure 4.3.

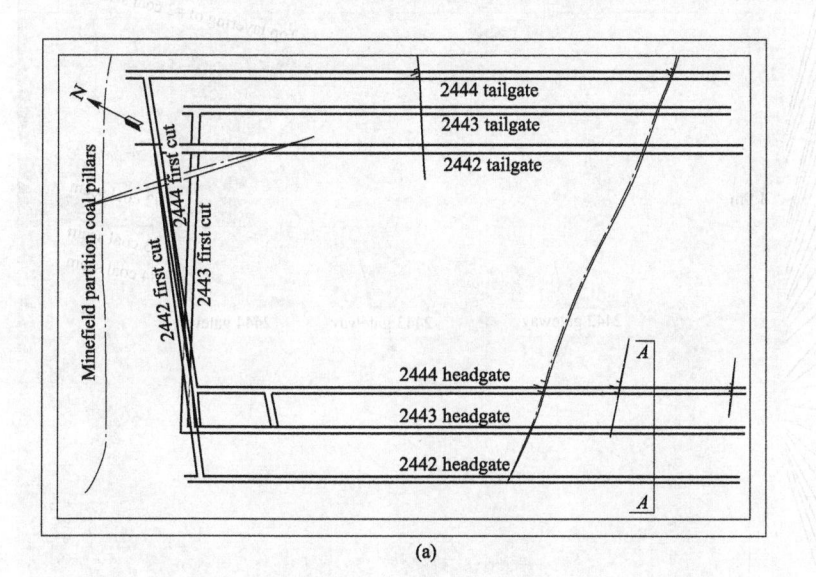

(a)

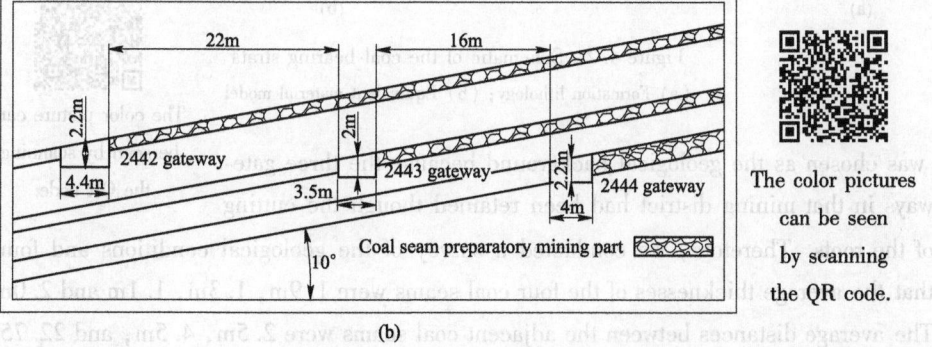

(b)

Figure 4.3　The gateway layout

(a) Plan view of part of the gateways of the #24mining district; (b) Section view of the study gateways of A-A

4.3.2 Equivalent Material Model

When choosing an appropriate geometric similarity ratio (C_L) before the experiment, the size of the test rig, the size of the excavation gateways, the thicknesses of the coal seam and stratum, and the end effect of the model all had to be considered. Firstly, the equivalent material model test was carried out using a rotating test rig with a size of 2m× 2m×0.3m. Secondly, the test involved the mining of three coal seams and the excavation and support of three gateways. Parts of the thicknesses of the coal seams and the stratum were small. The selection of the geometric similarity ratio had to facilitate laying the materials and excavating the gateway and the coal seams. In addition, the test had a large mining area of the model. The geometric similarity ratio had to be chosen to try to reduce the end effect of the model[20]. Considering the above factors, the geometric similarity ratio was determined as $C_L = L_M/L_H = 1/50$, where L_M is the size of the model and L_H is the size of the field. Furthermore, the bulk density similarity ratio was $C_\gamma = 0.667$[20] and the strength similarity ratio was $C_\sigma = C_L C_\gamma = 0.01334$.

The equivalent material employed fine river sand as an aggregate. Calcium carbonate and gypsum were used as the cementing material, and the size of the laying material was 2m×1.9m×0.3m within the test rig, as shown in Figure 4.2(b).

4.3.3 Load and Excavation of the Model

The 2442 headgate had a cover depth of 518m up to the ground level, and the equivalent material model could only provide the overburden thickness of 58m. Thus, the rest of the gravity of the strata needed load compensatory pressure, and this pressure could be calculated as 0.155MPa. Finally, five hydraulic jacks were employed for loading, and a steel plate was laid between the jacks and the model's surface to produce uniform pressure. The arrangement of jacks was shown in Figure 4.2(b). For safety reasons, the back plate was decorated on both sides of the jacks, which did not have contact with the model material. It mainly prevented the jacks and accessories from falling and injuring people during the loading process after rock failure.

In addition, it must be emphasized that the average spacing of 22.75m between the upward two coal seams was very large, and the mining of the protective coal seam of the top layering of the #2 coal seam did not badly damage the #2 coal seam and its adjacent stratum. Therefore, the experiment did not excavate the top layering of the #2 coal seam.

In the field, the roof of the 2442 headgate was reinforced by a bolt and cable, and the roofs of the 2443 and 2444 headgates were reinforced by a shed made with I-steel be-

· 50 · 4 Advancing Cutting Roof for Retaining Gateway With Near Horizontal and Close Coal Seam Groups

cause of the thin roof rock mass. To easily simulate the support and enable convenient operation, when the three gateways were excavated, a wood support was set to support the roof. Then a steel saw was used to cut the roof. The cut line deviation angle was $\gamma =$ 10°, the cut height (h) for the 2442 gateway was 69mm, which was the thickness of the two roof layers, and the cut heights (h) for the 2443 and 2444 gateways were the thicknesses of their roof layers.

The detailed excavating sequence during the test is shown in Figure 4.4. This sequence was consistent with the field excavating sequence.

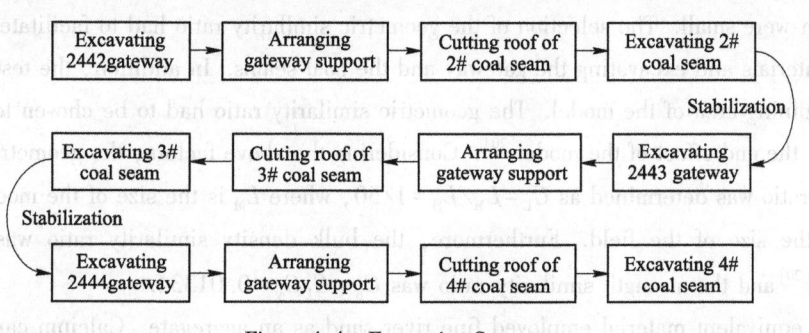

Figure 4.4 Excavating Sequence

4.4 Result of the Experiment

4.4.1 The Formation Process of the Roof Caving Structure

After cutting the roof, the roof above the gob caved in and the gateway of the surrounding rock had a stable state. The overall state of the caving roof and stratum is shown in Figure 4.5. The caving process and the cracks in the roof were recorded, as shown in Figure 4.6.

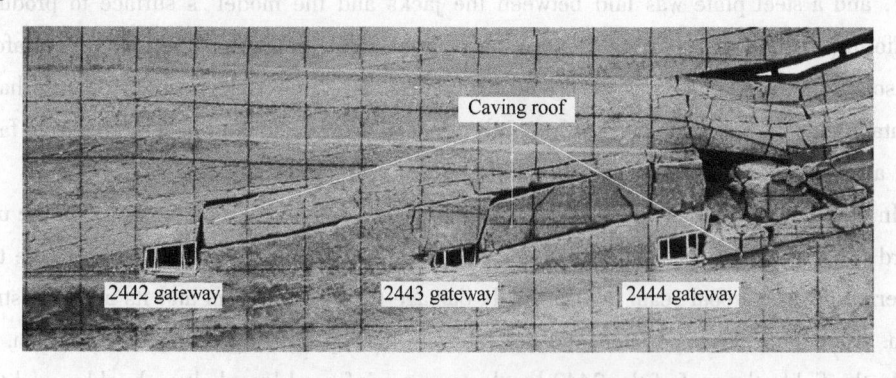

Figure 4.5 The overall state after excavation

4.4　Result of the Experiment　　　　· 51 ·

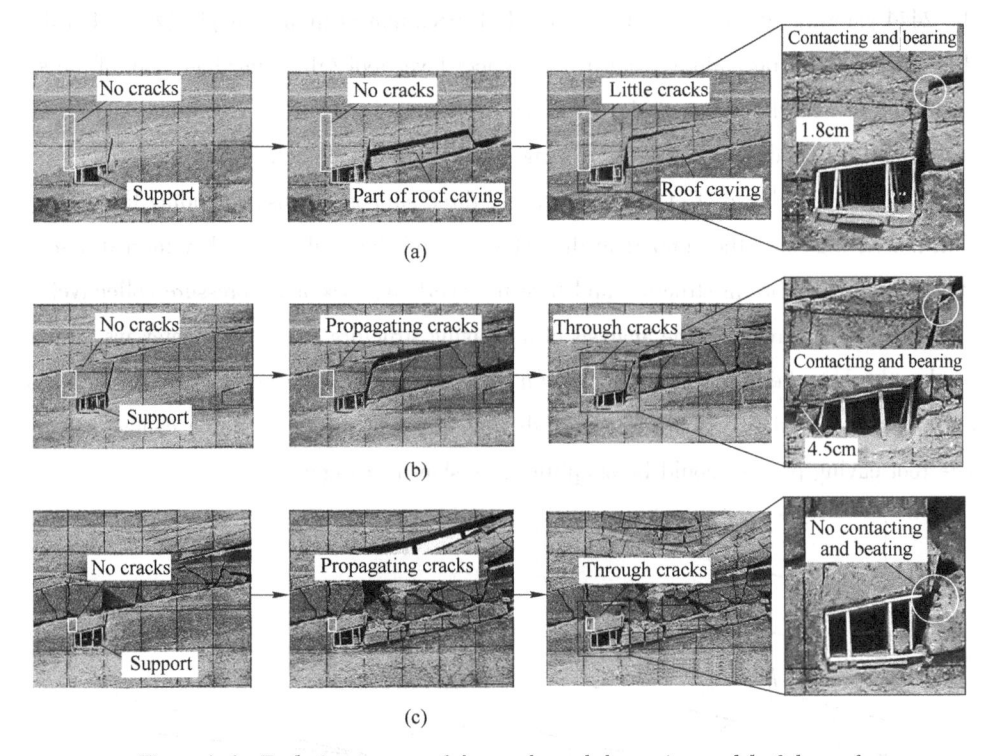

Figure 4. 6　Evolution process of the cracks and the caving model of the roof

(a) 2442 gateway; (b) 2443 gateway; (c) 2444 gateway

The following can be observed from Figure 4. 6:

(1) The coal side roof of the three gateways experienced no crack stages or longitudinal crack growth stages. After careful observation, it can be found that the cracks in the roofs of the 2443 and 2444 gateway eventually came through the roof, leading to the roof fracturing and sinking on the gob side under overburden pressure.

(2) The cutting roof of the 2442 gateway had two layers and the gob side roof experienced partial caving. The entire caving process (hereinafter referred to as contact) can be seen in Figure 4. 6(a). Moreover, the cutting roof of the 2443 gateway was single and thicker. The entire roof above the gob caved in, and it had a no contact and a contact process, as seen in Figure 4. 6(b). Furthermore, the entire cutting roof of the 2444 gateway above gob caved in but had no contact, as seen in Figure 4. 6(c). A further analysis of the results showed that the cut heights (h) of the 2442 and 2443 gateways were 3. 1 times and 4. 3 times the mining height (h_m). The cut height (h) was greater than the mining height (h_m), which led to a stability contact phenomenon for the caving roof. For

the 2444 gateway, the cut height (h) was 1.3 times that of mining height (h_m), but the difference was not big. More seriously, the upper layer roof (the immediate floor of the #3 coal seam) was a thin layer of clay rock that experienced serious damage when the #3 coal seam was excavated, leading to the caving roof having no contact.

(3) The no caving roof of the gateway had three support points, including the contact with the caving roof, the support in the gateway, and the coal side. They formed a self-stabilization equilibrium structure and bore the overlying rock mass pressure collectively.

Therefore, according to the analysis of the caving structure of the three gateways, it can be thought that when the roof cut height (h) was greater than the mining height (h_m) and the roof rock caved in without damage, the gateway roof formed a stable contact effect. The roof caving process could be simplified, as shown in Figure 4.7.

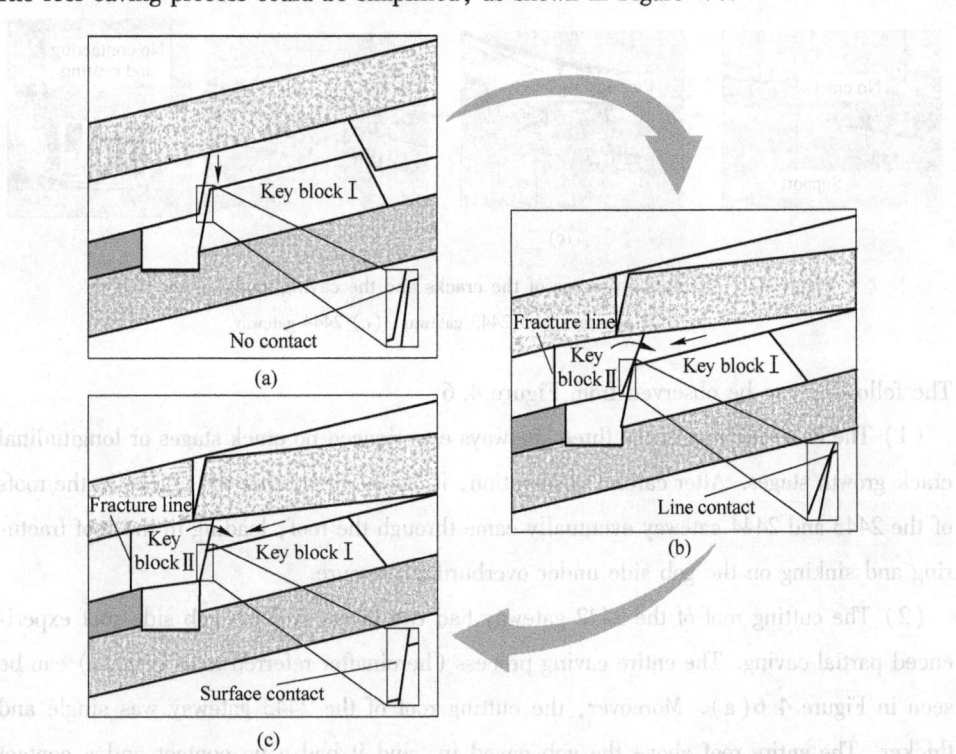

Figure 4.7 Caving structure model of the gateway roof that was cut

(a) Key block I fracture and caving; (b) Line contact of block I and key block II;

(c) Surface contact structure of block I and key block II

The gob side roof first caved in in the gravity direction when the roof achieved caving pace, but there was no contact. Then the roof on the other side of the caving roof (key block I) also experienced fracturing and caving, as shown in Figure 4.7(a). As the excavation advanced and the stress was adjusted, the gateway roof fractured and a vertical fracture line appeared above the shallow coal side and formed key block II. This rotated the point to the gob side along the lower endpoint of the fracture line and had a line contact with key block I, as shown in Figure 4.7(b). With the increase of the roof deformation and extrusion pressure, the contact area also increased and formed a further stable surface contact structure: the 'load-bearing structure' equilibrium model, as shown in Figure 4.7(c). In addition, it can be found from the equivalent material simulation experiment that the coal and rock mass had serious deformation and failure, but key block II still had contact with the fracture roof above the coal side and produced a mechanical effect.

4.4.2 The Mechanical Structure Model of the Caving Roof

4.4.2.1 The Analysis of the Balance Control Force of the Model

The stability of the key block II controlled the equilibrium state of the 'self-stabilization bearing structure'. In order to carry out the force analysis of key block II, the following assumptions and regulations were used[21]:

(1) It was assumed that key block II was a rigid body that would not deform when forced.

(2) The rotary angle of the key block II point to the gob was very small. Thus, it was assumed that key block II did not rotate after fracturing to simplify the calculation. In addition, the lower endpoint of the fracture line of key block II presented a mechanical effect, and it was assumed that it was subjected to a horizontal force.

(3) It was assumed that there was a surface contact between key block I and key block II, and that a mechanical effect was produced on the contact surface.

(4) It was assumed that the coal wall provided a supporting force to the roof, and that the force was equal everywhere on the acting surface.

According to the above assumptions, the force analysis of key block II was carried out, as shown in Figure 4.8.

To gain an in-depth understanding to the above model, the mechanical equilibrium equations: $\sum F_h$ for the horizontal direction, $\sum F_v$ for the vertical direction, the moment $\sum M_s$ for point S and the moment $\sum M_t$ for point T, were established, given by:

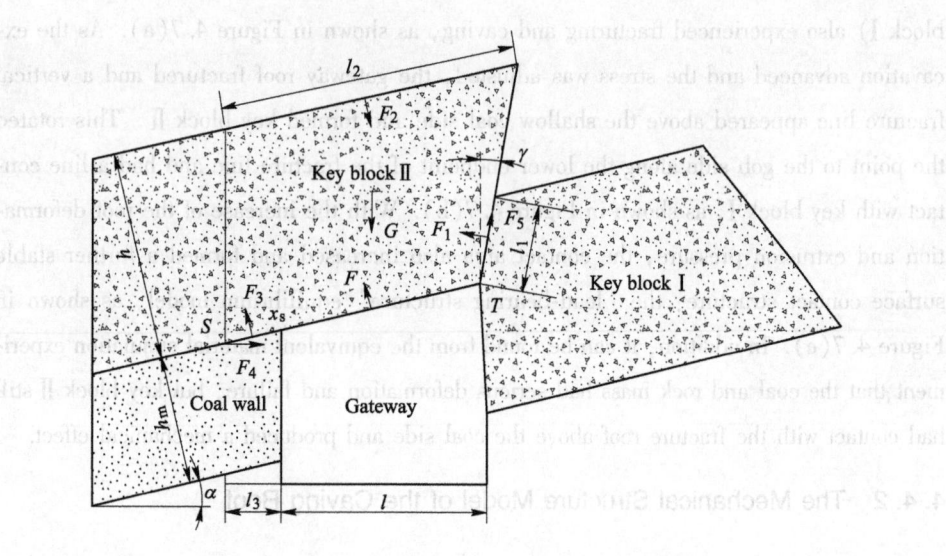

Figure 4.8 The force analysis of key block II piece

$$
\begin{cases}
F\sin\alpha + F_1\cos\gamma + F_3\sin\alpha - F_2\sin\alpha - F_4 - F_5\sin\gamma = 0, \sum F_h = 0 \\
F\cos\alpha + F_1\sin\gamma + F_3\cos\alpha - F_2\cos\alpha + F_5\cos\gamma - G = 0, \sum F_v = 0; \\
G\sqrt{\dfrac{l_2}{2}\dfrac{x_S+l}{2}}\cos\alpha + F_2\left(\dfrac{l_2}{2} + h\tan\alpha\right) - F_1\left[\dfrac{l_1}{2} + \dfrac{l+l_3}{\cos\alpha}\sin(\alpha+\gamma)\right] \\
\quad - F\left(\dfrac{l}{2\cos\alpha} + x_S\right) - F_3\dfrac{x_S}{2} - F_5\dfrac{l+l_3}{\cos\alpha}\cos(\alpha+\gamma) = 0, \sum M_S = 0 \\
\quad - G\left(x_S + l - \sqrt{\dfrac{l_2}{2}\dfrac{x_S+l}{2}}\right)\cos\alpha - F_2\left(x_S + l - \dfrac{l_2}{2} - h\tan\alpha\right) - F_1\dfrac{l_1}{2} + \\
F\dfrac{l}{2\cos\alpha} + F_3\left(\dfrac{x_S}{2} + \dfrac{l}{\cos\alpha}\right) - F_4(l+l_3)\tan\alpha = 0, \sum M_T = 0
\end{cases}
$$

$$(4.1)$$

Where　F——the roof support force;

　　　F_1——the normal force on the contact surface of the two key blocks;

　　　F_2——the force from the gateway roof;

　　　F_3——the coal wall support force;

　　　F_4——the horizontal force of the roof acting on key block II;

　　　F_5——the friction on the contact surface of the two key blocks;

　　　G——the gravity of key block II;

l_1——the contact distance of the two key blocks;

α——the coal seam dip angle;

x_S——the distance from the roof break point to the coal wall;

l_3——the horizontal distance from the roof break point to the coal wall;

l——the gateway width.

Some variables [in Eq. (4. 1)] are defined as follows:

$$l_3 = x_S \cos\alpha \tag{4.2}$$

$$l_1 = \frac{h - h_m}{\cos(\gamma + \alpha)} \tag{4.3}$$

$$G = \frac{1}{2}\left(l_2 + \frac{l + l_3}{\cos\alpha}\right) h\lambda \tag{4.4}$$

$$F_5 = F_1 w_1 \tag{4.5}$$

Where λ——the bulk density of the gateway roof;

w_1——the friction coefficient of the contact surface of the two key blocks.

Furthermore, the force of F_2 was calculated using Terzaghi's principle[21] as:

$$F_2 = \frac{\lambda_1 l_2^2}{2A_x \tan\varphi} \cos\alpha \tag{4.6}$$

Where λ_1——the average bulk density of the loading layer;

φ——the average internal friction angle of the loading layer;

A_x——the lateral pressure coefficient, given by:

$$A_x = 1 - \sin\varphi \tag{4.7}$$

In addition, l_2 is the upper surface length of key block Ⅱ, as shown in:

$$l_2 = h\tan(\alpha + \gamma) - h\tan\alpha + \frac{l + l_3}{\cos\alpha} \tag{4.8}$$

4. 4. 2. 2 Solving the Mechanical Structure Model

Through the above analysis, it can be found that the roof support force of F, the normal force on the contact surface of F_1, and the coal wall support force of F_3 played an important role in controlling key block Ⅱ. Therefore, the relationship between the three balance control forces, the cut height (h), and the cut line deviation angle (γ) was calculated. Firstly, it must be emphasized that the geological conditions and parameters (Listed in Table 4. 1) of the 2442 gateway were chosen as the calculation example, and the following two steps were executed:

（1）The cut line deviation angle of $\gamma = 10°$ was kept constant and the cut height (h) increased from 2m to 5m in steps of 0.5m. The final results are shown in Figure 4.9(a).

（2）The cut height of $h = 3.9$m was kept constant and the cut line deviation angle (γ) increased from $0°$ to $30°$ in steps of $5°$. The final results are shown in Figure 4.9(b).

Table 4.1 Values of the calculation parameters

Parameter	Value	Parameter	Value	Parameter	Value
α	$10°$	λ	$25.9 \times 10^3 \text{N/m}^3$	x_S	0.9
h	3.9m	λ_1	$18 \times 10^3 \text{N/m}^3$	w_1	0.5
h_m	1.1m	γ	$10°$		
φ	$40°$	l	4.4		

Figure 4.9 shows the following results:

（1）Figure 4.9(a) shows that with the increase of the cut height, the roof support force F and the coal wall support force F_3 increased approximately linearly. In contrast, the support force F_1 on the contact surface first decreased and then increased. Moreover, it can be found that the rangeabilities of F and F_3 were significantly higher than that of F_1 was, indicating that the cut height had a greater influence on F and F_3 than that of F_1.

（2）Figure 4.9(b) shows that with the increase of the cut line deviation angle, the roof support force F increased approximately linearly. In contrast, the support force F_1 on the contact surface coal wall and the support force F_3 first decreased and then increased. Moreover, it can be found that the rangeability of F was significantly higher than those of F_1 and F_3 were.

Previous studies[3,5,22,23] have shown that a low support force is usually selected to control the surrounding rock of a gateway and to further determine the parameters of the cut height and the cut line deviation angle based on the selected support force. As shown in the above analysis, the support forces F, F_3, and F_1 were the most important forces for controlling the gateway roof. Hence, their relationship should be further studied.

It can be seen that when the values of the cut height and the cut line deviation angle were the smallest. Additionally, the support force F was the smallest (shown in Figure 4.9), meaning that a worker could provide a low support force F to control the roof.

4. 4   Result of the Experiment   · 57 ·

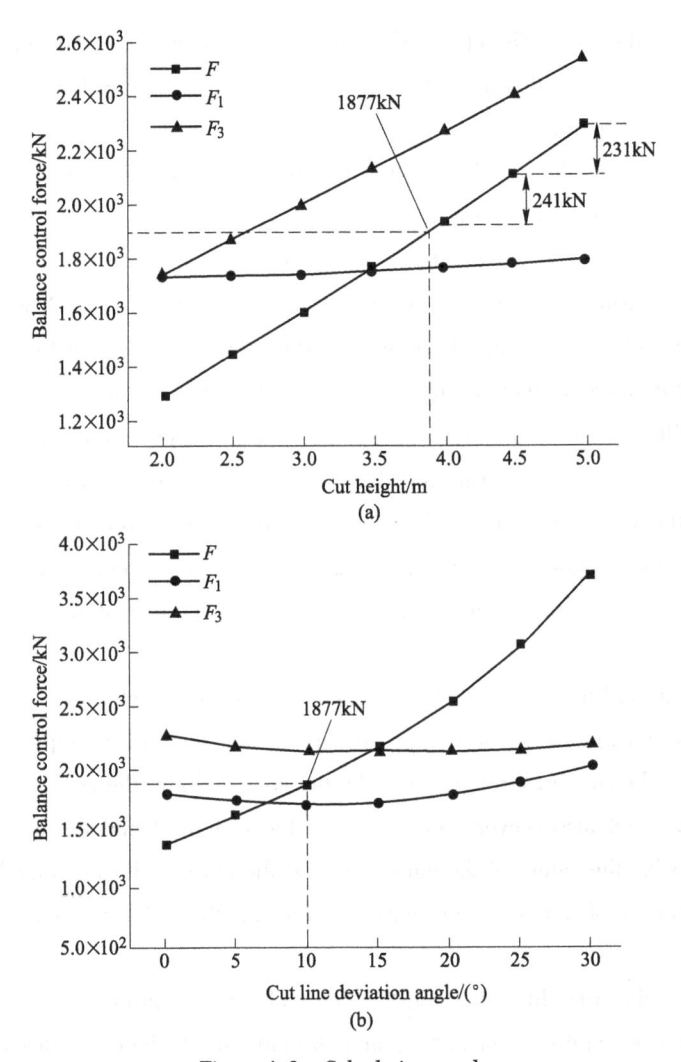

Figure 4. 9   Calculation results

(a) Balance control force with cut height; (b) Balance control force with cut line deviation angle

However, the support forces F_3 and F_1 were greater than the support force F at that moment. Moreover, it was realized there was the serious issue of the coal wall and the contact surface of the two key blocks undergoing some deformation, and the coal wall and the key blocks had the hidden dangers of failure and loss of stability when the support forces F_3 and F_1 were larger. In addition, it can also be found that even if the coal wall had no deformation, the roof of key block Ⅱ was a deformable body whose parts in the coal wall with a width of l_3 and above the roof with a width of l had a different degree of deformation when the values of the support forces of F and F_3 had a big gap. This led to more damage to the roof. Therefore, the smallest values of the cut height and the cut line deviation angle were determined based on the smallest support force F that was unfavorable to the surrounding rock control.

Since the smallest values of the cut height and cut line deviation angle posed a safety hazard, increasing their value was imperative. Further analysis found that with the increase of the value of the cut height, the support force F increased, but the support force F_3 had a substantial increase, as shown in Figure 4.9(a), which was more unfavorable to the coal wall stability. In addition, when increasing the cut line deviation angle, the support force F had a sharp increase. The support forces F_1 and F_3 varied weakly but still maintained the high value of a 2×10^3 kN magnitude compared to the values in Figure 4.9(a), which were also bad for the surrounding rock control. However, according to the values in Figure 4.9(b), when cut line deviation angle increased to a certain value, the support forces of F, F_1, and F_3 were close to each other. This could satisfy the coordinate deformation of the different parts of the roof rock mass as far as possible.

Therefore, according to the above analysis, the selection of the values of the cut height and the cut line deviation angle had conform to a certain principle that it should not only utilize the support force provided by the coal wall and the contact surface of the two key blocks, but also prevent the failure of the coal wall and the contact surface. Thus, in choosing the values of the cut height and the cut line deviation angle for advancing the cutting roof for gob-side entry retaining, the following steps had to be adhered to:

(1) In general, to facilitate the construction technology control, the values of the cut height were integral multiples of 0.5m, and the cut line deviation angles were integral multiple of 5° when the roof was thick.

(2) In practice, the selection of cut height was influenced by the roof layered properties and the spacing between the coal seams. For example, the roof of the 2442 gateway had more layers and the thickness of each layer was thinner, so the cut height had to be larger than the mining height and it was necessary to try to cut several layers together at the same time. In addition, the roofs of the 2443 and 2444 gateway were located below the gob, and they were cut one time. However, the cut line deviation angle was almost controlled by a human. Therefore, the cut height had to be determined first.

(3) The cut line deviation angle was selected according to the above model, and the values of the support forces of F, F_1, and F_3 were made very level to coordinating sup-

port for key block II.

(4) After determining the two parameters, the coal wall had to be reinforced to increase its ability to resist deformation in the field.

According to the above principles, the roof cut height of 3.9m for the 2442 gateway was more appropriate. Moreover, the model in Figure 4.9(b) shows that the cut line deviation angles of 10° or 15° could be selected, but the latter was the best.

4.5 Discussion

An equivalent material simulation experiment was conducted in order to explain other phenomena in the process of the cutting roof for gob-side entry retaining, especially because it was a surprise to discover this (self-stabilization bearing structure) equilibrium model. In addition, the purpose of this study is to expound the discovery process of the model and its mechanical structure. Thus, many parameters of the equivalent material model, such as the mixed ratio of the materials and strength of materials, were not listed in the study. However, they can be found in references[24-26]. However, it can be noted that the value of x_S was 0.8m, which was calculated according to the geometric similarity ratio of the 1/50 reference of the experimental data in Figure 4.6(a). There is no definite answer at present for whether the parameter of x_S was high precision. In addition, the geological condition and excavation impact extent of each gateway were varied, like the above three gateways. Therefore, determining how to acquire the theoretical value of x_S still needs further study.

Moreover, it can be found that the proposed 'self-stabilization bearing structure' equilibrium model is equivalent to the hypothesis of a stope 'articulated beam' proposed by H. Kuznetsov of the former Soviet Union using an equivalent material model experiment[21]. Furthermore, because the model could take into account the many factors, such as faults and precise physical parameters, present in the actual strata, the analysis results may be considered as some kind of indication for mining engineering practice. However, in reality, this model took full advantage of the strength of the caved in hard roof rock mass to support itself, and could improve the stability of the roof. After the above theoretical analyses, it can be found that if gateway had a better industrial design for advancing cutting roof for gob-side entry retaining, and there were appropriate cables in the roof, bolts in the coal wall, and a single hydraulic prop in gateway, as shown in Figure 4.10[27-32], the roof stability would be significantly improved, and the coalmine would be more conducive to safe and efficient production with a hard roof stratum.

· 60 · 4 Advancing Cutting Roof for Retaining Gateway With Near Horizontal and Close Coal Seam Groups

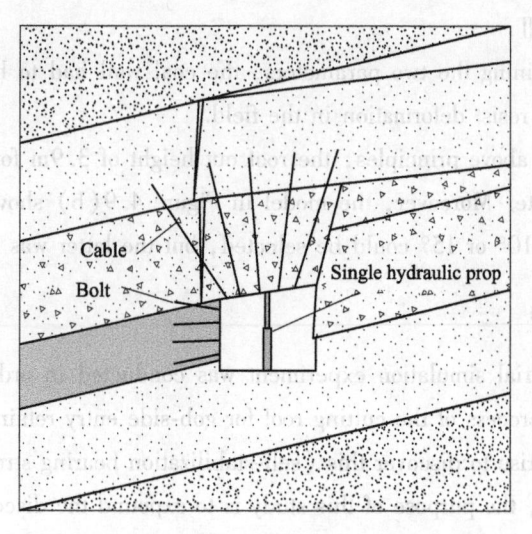

Figure 4. 10 Support diagram of a gateway with a cable, bolts, and a single hydraulic prop

4. 6 Chapter Summary

An advancing cutting roof for gob-side entry retaining has many advantages compared with the traditional method of constructing an artificial filling wall, and it will be used in more and more gateways in coalmines. To obtain more in-depth information about this technology, an equivalent material model experiment with a plane stress state was carried out based on the complex geological conditions of a flat seam and a close distance coal seam group for a Baijiao coalmine. Based on the three experimental gateways, it could be found that the cut line deviation angle and the cut height were two important parameters. When the cut line deviation angle was unequal to zero, the cut height was greater than the mining height, and the caving roof rock was hard without damage, a 'self-stabilization bearing structure' equilibrium model was ultimately formed, which was conducive to the stability of the gateway roof. The model showed that its stability was mainly controlled by two fractured and caved roof blocks, key blocks I and II. Furthermore, in order to determine the optimal parameters of the cut height and the cut line deviation angle for the cutting roof of a retaining gateway, mechanical analysis of the model was performed assuming that the key blocks were made of rigid material. Additionally, the relationship between the roof balance control force, the cut height, and the cut line deviation angle was solved. After an in-depth analysis of the result calculated by the mechanical model, it was found that the selection of the values of the cut height and the cut line deviation angle had to conform to a certain principle that it should not only utilize the support force provided by the coal wall and the contact surface of the two key blocks but also prevent the failure of the coal wall and the contact surface. Finally, it is predicted that using this structure with a perfect industrial design will be more conducive to safe and efficient production for a coalmine with a hard roof stratum.

References

[1] Wu B W, Wang X Y, Bai J B, et al. Study on crack evolution mechanism of roadside backfill body in gob-side entry retaining based on UDEC trigon model[J]. Rock Mech Rock Eng, 2019, 52(9): 3385-3399.

[2] Sun X M, Liu Y T, Wang J W, et al. Study on three-dimensional stress field of god-side entry retaining by roof cutting without pillar under near-group coal seam mining[J]. Processes, 2019B, 7(9).

[3] Zhang X Y, Chen L, Gao Y B, et al. Study of an innovative approach of roof presplitting for gob-side entry retaining in longwall coal mining[J]. Energies, 2019B, 12(17).

[4] He M C, Ma X G, Y B. Analysis of strata behavior process characteristics of gob-side entry retaining with roof cutting and pressure releasing based on composite roof structure [J]. Shock Vib, 2019.

[5] Zhang G, He M, Yu X, et al. Research on the technique of no-pillar mining with gob-side entry formed by advanced roof caving in the protective seam in baijiao coal mine[J]. J Min Saf Eng, 2011, 28(4): 511-516.

[6] Sun X, Liu X, Liang G, et al. Key parameters of gob-side entry retaining formed by roof cut and pressure releasing in thin coal seams[J]. Chin J Rock Mech Eng, 2014, 33(7): 1449-1456.

[7] Zhang G, Xu Y, Ge P. Research on cut gob-side entry retaining in thin coal seam of Tangshan ditch, Chin[J]. J Rock Mech Eng, 2016, 35(7): 1397-1406.

[8] Liu Y, Li W, Huang X. The application of roof cutting and pressure relief blasting technology in gob-side entry retaining[J]. Saf Coal Mine, 2014, 45(6): 132-135.

[9] Zhang Z Z, Wang W J, Li S Q, et al. An innovative approach for gob-side entry retaining with thick and hard roof: a case study[J]. The Vjes, 2018, 25(4): 1028-1036.

[10] Liu H, Dai J, Jiang J Q, et al. Analysis of overburden structure and pressure-relief effect of hard roof blasting and cutting[J]. Adv Civ Eng, 2019.

[11] Wang P, Jiang J, Zhang P, et al. Breaking process and mining stress evolution characteristics of a high-position hard and thick stratum[J]. Int J Min Sci Technol, 2016, 26(4): 563-569.

[12] Ta D S, Cao S G, Steyl G, et al. Prediction of groundwater inflow into an iron mine: a case study of the Thach Khe Iron Mine, Vietnam [J]. Mine Water Environ., 2019, 38(2): 310-324.

[13] Liu X S, Ning J G, Tan Y L, et al. Coordinated supporting method of gob-side entry retaining in coal mines and a case study with hard roof[J]. Geomech Eng, 2018, 15(6): 1173-1182.

[14] Ma Q, Tan Y L, Zhao Z H, et al. Roadside support schemes numerical simulation and field monitoring of gob-side entry retaining in soft floor and hard roof[J]. Arab J Geosci, 2018, 11(18): 13.

[15] Hu J Z, Zhang X Y, Gao Y B, et al. Directional presplit blasting in an innovative no-pillar mining approach[J]. J Geophys Eng, 2019, 16(5): 875-893.

[16] Guo Q, Guo G, Lv X, et al. Strata movement and surface subsidence prediction model of dense solid backfilling mining[J]. Environ Earth Sci, 2016, 75(21): 1426.

[17] Zhou S, Wu K, Zhou D, et al. Experimental study on displacement field of strata overlying goaf with sloping coal seam[J]. Geotech Geol Eng, 2016, 34(6): 1847-1856.

[18] Alencar A S, Galindo R A, Melentijevic S. Bearing capacity of foundation on rock mass depending on footing shape and interface roughness[J], Geomech Eng, 2019, 18(4): 391-406.

[19] Yang D W, Ma Z G, Qi F Z, et al. Optimization study on roof break direction of gob-side entry retaining by roof break and filling in thick-layer soft rock layer[J]. Geomech Eng, 2017, 13(2): 195-215.

[20] Luo F, Yang B, Hao B, et al. Mechanical properties of similar material under uniaxial compression and the strength error sources[J]. J Min Saf Eng, 2013, 30(1): 93-99.

[21] Qian M, Shi P, Xu J. Ground Pressure and Strata Control[M]. Xuzhou: China University of Mining and Technology Press, 2012.

[22] Song Z, Konietzky H. A particle-based numerical investigation on longwall top coal caving mining[J]. Arab J Geosci, 2019, 12: 556.

[23] Song Z, Wei W, Zhang J. Numerical investigation of effect of particle shape on isolated extracted zone (IEZ) in block caving[J]. Geo Ssci, 2018, 11: 310.

[24] Tang J. Atypical dynamic phenomena of mines and assessment method[D]. Chongqing: Chongqing University, 2004.

[25] Li Z. Study on pressure control of soft rock laneway at great depth under complex stress condition [D]. Chongqing: Chongqing University, 2006.

[26] Yang Q. Roof control and application of comprehensive technology of fire prevention and extinguishing of residual coal compound mining of baijiao coal mine[D]. Xi'an: Xi'an University of science and technology, 2013.

[27] Kang Y S, Liu Q S, Xi H L, et al. Improved compound support system for coal mine tunnels in densely faulted zones: A case study of China's Huainan coal field[J]. Eng Geol, 2018, 240: 10-20.

[28] Chen Z Q, He C, Xu G W, et al. Supporting mechanism and mechanical behavior of a double primary support method for tunnels in broken phyllite under high geo-stress: a case study[J]. Bull Eng Geol Environ, 2019, 78(7): 5253-5267.

[29] Yang S Q, Chen M, Jing H W, et al. A case study on large deformation failure mechanism of deep soft rock roadway in Xin'An coal mine, China[J]. Eng Geol, 2017, 217: 89-10.

[30] Wang C L, Li G Y, Gao A S, et al. Optimal pre-corditioning and support designs of floor heave in deep roadways[J]. Geomech Eng, 2018, 14(5): 429-437.

[31] Aksoy O C, Uyar G G, Utku S, et al. A new integrated method to design of rock structures[J]. Geomech Eng, 2019, 18(4): 339-352.

[32] Oh J, Moon J, Ganbulat I, et al. Design of initial support required for excavation of underground cavern and shaft from numerical analysis[J]. Geomech Eng, 2019, 17(16): 573-581.

5 Natural Filling and Systematic Roof Control Technology for Retaining Gateway in Steep Coal Seams

5. 1 Introduction

Previous studies showed that the dip of the coal seam was the most important geological factor[1, 2] because the dip angle affected the difficulty of implementing GER. Thus, because of the limitation of the technique for controlling the surrounding rock, the vast majority of GER cases have focused on flat and gently inclined coal seams with dips of less than 25°[3-6]. In addition, no pillar is needed for the retained entry, but rather a filled wall on the gob side is constructed, usually by one of five materials: high-water material (HWM), concrete, paste, blocks, or waste pack[5].

Gradually, people found that caved in waste rock would automatically slide or roll to the low end of the working face in the gob of a steep coal seam with a dip of 35° to 55° according to the natural repose angle of the caved waste rock[2]. Then, they realized that the caved waste rock could be used as filling material in steep coal seams. In addition, if the caved waste rock can be used up gob-side filling material, it will not only decrease the cost and labor intensity, but will also meet the requirements of green coal mining[7]. In addition, experiments conducted on the use of broken gangue as the gob filling material[8,9], found that it had good physical properties, such as a high compressive strength and a greater residual strength than that of broken shale and sandstone, and a high compressibility. So it fully meets the supportive resistance, rotation, and sinking adaptation requirements of large roof structures, making it an ideal gob side filling material[10,11].

However, GER in steep coal seams is closely connected to the movement of the roof. Many studies have been conducted on the mechanical structure, stress distribution, deformation, and failure process of steep rooves by 'clamped beam' and 'simply support beam' theory based on various conditions and mining characteristics[12]. It was concluded that the main roof would form 'voussoir beam' structures, and an immediate roof located at the middle-lower part of the working face would probably form a small scale 'voussoir arch' structure due to the waste filling effect[13]. In addition, it was found that a steep coal seam had asymmetric mechanical characteristics along the dip direction. Roof caving began in the middle-upper part of the working face and continued into the upper strata and the lower part of the working face. Finally, the caving range could extend to the tailgate with the working face advancing[14,15]. In addition, based on previous studies, the failure mechanism of the gateway in a steep coal seam was analyzed. As the dip angle increases, the gateway's failure location will mainly be concentrated on the ribs, the roof's upper corner, and the floor's lower corner due to roof deformation and floor sliding[16-18].

Though we have a great deal of knowledge concerning GER, the utilization of GER in steep coal seams has not been extensively developed to date. In recent years, as mining science progressed, GER technology in steep coal seams has begun to be considered and accepted. The use of GER has been tested in several steep coal seams, and a certain amount of success has been obtained [2,19-21]. But even so, many technical problems still exist and have not been systematically analyzed. Therefore, we conducted our study to further develop our understanding. First, the steep coal seams were investigated. Second, the characteristics of the caved in rooves and the stress evolution were determined using three-dimensional distinct element code[22]. Third, the rock blocking device and grouting reinforcement method of the natural filling technology and the hydraulic support for strengthen and supporting the gateway were developed. Finally, the gateway cross section and its driving support were designed taking into consideration the earlier findings.

5.2　Investigation of Steep Coal Seams

The world's major coal-producing countries, e. g. , Russia, the UK, Germany, and Poland, have a large proportion of their reserves in steep coal seams. In China, these reserves account for 14. 05% of the total coal reserves, and in some provinces, it is an even larger proportion[23]. There are 4 main coalfields in the southwestern part of Sichuan Province, China[24].

The coal-bearing strata contain a large number of tectonic structures, resulting in diverse and complex geological conditions. To obtain detailed dip angles for the area, 45 workable coal seams within the 4 coalfields were analyzed. It was found that 56% of the investigated coal seams (25 coal seams) have dips of more than 35°. In addition, these steep coal seams accounts for about 20% of the proven reserves and 10% of the coal output of China's coal seams, and more than 50% of these contain high quality coking coal and anthracite[25]. Thus,the coal recovery rate must be improved,and the use of GER technology in steep coal seams is critical and necessary.

5.3　Analysis of the Advantage of the Use of GER in Steep Coal Seams

5.3.1　Analysis of the Characteristics of Roof Cave-Ins

It is well-known' that after the longwall mining working face advances, the roof will cave-in. The characteristics of this caving in steep coal seams are shown in Figure 5.1(a), assuming the caved waste rock is blocked on the gob side[2,13,15,26].

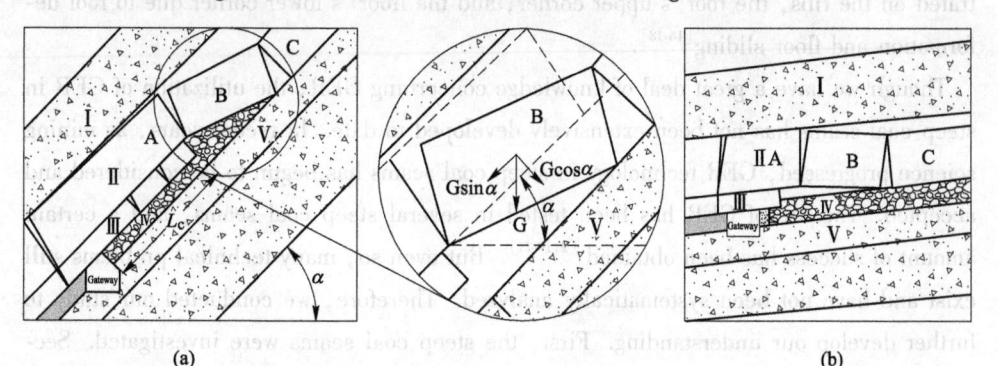

(a)　　　　　　　　　　　　　　　　　(b)

Figure 5.1　Characteristics of roof cave-ins

(a) Characteristics of roof cave-ins in steep coal seams;

(b) Characteristics of roof cave-ins in flat and gently inclined coal seams

I —Overburden strata; II —Main roof; III —Immediate roof; IV —Caved waste rock; V —Immediate floor

The characteristics of roof cave-ins, mainly involve the main roof caving structure, the filling characteristic, and the dynamic effect. In this study, they were analyzed as follows:

(1) The main roof caving structure. The characteristics of the roof's failure and movement exhibit an obvious asymmetry. In the inclined direction, the middle-upper part of the roof caved first and periodic weighting, thus resulting in an enlarged caving space, and greatly rotated deformation of the main roof, which fractured and formed 'voussoir beam' structures.

(2) The filling characteristic. After the middle-upper part of the immediate roof caved, the caved waste rock rolled or slid toward the lower working face, and finally amassed there. The inclination length of the waste rock filling and supporting part is L_c, as seen in[13]:

$$L_c = \frac{\bar{h}k_i}{\bar{h} + h}L - 0.5(\bar{h} + h)\cot(\alpha - \beta) \tag{5.1}$$

Where　L——the inclination length of the working face, m;

k_i——the initial bulking factor of the immediate roof;

$\bar{h}$——the average caving height of the immediate roof, m;

h——the mining height, m;

α——the dip angle of the coal seam, (°);

β——the natural repose angle of the caved waste rock, $\beta = 35°$.

Because of the filling effect of the long part, it is a long distance from the bending, rotating, and breaking location of the main roof to the gateway, which lowers the influence of the cave-in on the gateway's stability compared to that of flat and gently inclined coal seams, as shown in Figure 5.1(b), making steep coal seams more advantageous to the stability of the gateway.

(3) The dynamic effect. The caving of the main roof has a dynamic effect on the gateway, and the long filling distance of the waste rock greatly weakens its dynamic damage effect compared to that of the flat and gently inclined coal seams, as shown in Figure 5.2(b). In addition, with increasing dip angle, the gravity component ($G\cos\alpha$) of the key block (B), as shown in Figure 5.1(a), will decrease after it loses stability, which can also reduce the dynamic effect.

This demonstrates that the characteristics of roof cave-ins in steep coal seams are advantageous to GER.

5.3.2 Analysis of the Stress Redistribution

In the case of stress redistribution caused by roof cave-ins, the secondary stress has a large impact on the stability and integrity of the gateway surrounding the coal and rock masses.

5.3.2.1 Numerical Model

To study the stress redistribution, one coal mine with steep coal seams was investigated, and a 3DEC[22] model was employed for plane-strain conditions. The dimension of the model was 170m×5m×175m, in the x, y and z direction, respectively, and it included the coal seams and rock strata, with a total of 7 layers, as shown in Figure 5.2(a), in accordance with the geological conditions of the investigated mine. The Mohr-Coulomb yield criterion was used for the strata and the Coulomb slip model was used for the contacts. The mechanical and physical properties of all the layers and the contacts between the layers are described in Deng and Wang[19] and Gao et al. [27], respectively. All of the side boundaries were roller-constrained, and the bottom was fixed. The upper boundary was subjected to a uniformly distributed vertical stress, for detailed see Deng and Wang[19].

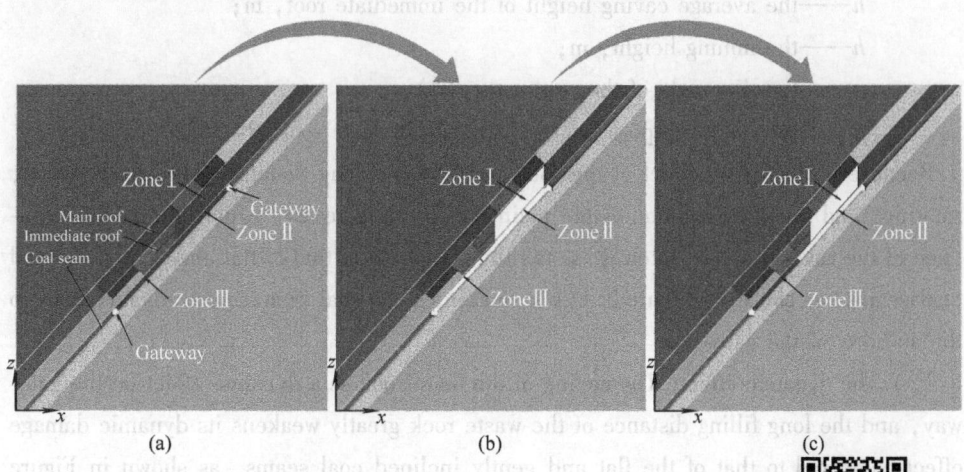

(a)   (b)   (c)

Figure 5. 2   Numerical model of the roof cave-in of a steep coal seam
(a) Numerical model; (b) 'Excavated' step; (c) 'Fill' step

First, we used the model to perform an initial equilibrium calcula-  The color pictures can be seen by scanning the QR code.
tion. Then, the headgate and tailgate were excavated and the model
was run until it reached equilibrium again. Finally, the roof was re-
moved in 'Zone I', 'Zone II', and 'Zone III' of the coal seam
using the 'excavate' command, as shown in Figure 5. 2(b). Then the gob backfill zone
of 'Zone III' was filled with gob filling material using the 'fill' command, as shown in
Figure 5. 2(c), and the model continued to calculate.

5. 3. 2. 2 Gob Filling Material

The gob filling material (caved rock) is a strain-stiffening material. After the initial
high compaction, the material will be stiffer and the modulus of the compacted aggregate
will increase. The double-yield model, which is intended to represent the materials in
which there may be significant irreversible compaction in addition to shear yielding, is
available in 3DEC. Hence, the double-yield (DY) model has been widely used to simulate gob material in many studies[28-30]. Salamon's model is valid to simulate the stress-
strain relationship of cave-in materials, as shown in as follows:

$$\sigma = \frac{E_0 \varepsilon}{1 - \dfrac{\varepsilon}{\varepsilon_m}} \tag{5.2}$$

Where σ——the uniaxial stress applied to the gob material while the material is

5.3 Analysis of the Advantage of the Use of GER in Steep Coal Seams

rigidly confined laterally;

E_0——the initial tangential modulus of the material;

ε——the strain occurring under the applied stress;

ε_m——the maximum possible strain;

E_0 can be obtained from Equation (5.3)[31], given by:

$$E_0 = \frac{10.39\sigma_c^{1.042}}{k_i^{7.7}} \tag{5.3}$$

Where σ_c——the compressive strength of the caved rock piece $\sigma_c = 30$MPa;

k_i——the initial bulking factor of the caved rock.

For the determination of ε_m, some studies have used $\varepsilon_m = (k_i - 1)/k_i$ and $k_i = (h_m + h_n)/h_n$ to calculate in flat seam. h_m is the mining height, and h_n is the roof caving height. However, the situation is more complex in a steep coal seam, as the caved rock will roll or slide to the lower part of the working face where only existed the mining height and no caving height. In addition, the equation for ε_m was obtained assuming only that the maximum displacement of the caved rock is the mining height, which cannot be reached. Thus, the equations for ε_m and k_i cannot be used in the case of steep coal seams. In reality, the caved rock has a compression limit, which can be expressed by the residual bulking factor (k_r), and the maximum displacement of the caved rock can be controlled by k_i and k_r. Hence, ε_m can be determined from Equation (5.4):

$$\varepsilon_m = \frac{h - \dfrac{h}{k_i}k_r}{h} = 1 - \frac{k_r}{k_i} \tag{5.4}$$

Site investigations in coal mines show that the bulking factor of coal rocks is 1.1 to 1.5[28]. During the rolling or sliding process of waste rock, the initial bulking factor can be defined as $k_i = 1.5$, and the residual bulking factor can be defined as $k_r = 1.1$. Thus, the maximum possible strain (ε_m) of the gob material and the initial tangent modulus (E_0) can be calculated as 0.36 and 15.8 MPa, respectively. The stress-strain curve for Equation (5.2) had been plotted and illustrated in Figure 5.3(a).

To determine the gob's material parameters, a simple zone was defined with dimensions of 1m×1m×1m, as shown in Figure 5.3(a)[30]. Loading was simulated by applying a velocity to the top of a model with confined sides. The strain-stiffening curve from Equation (5.2) for the given variables was fitted using an iterative change in the bulk and shear moduli and the angle of friction of the gob filling material, as shown in Figure 5.3(a). The final parameters obtained and used in the gob DY model are listed in Table 5.1.

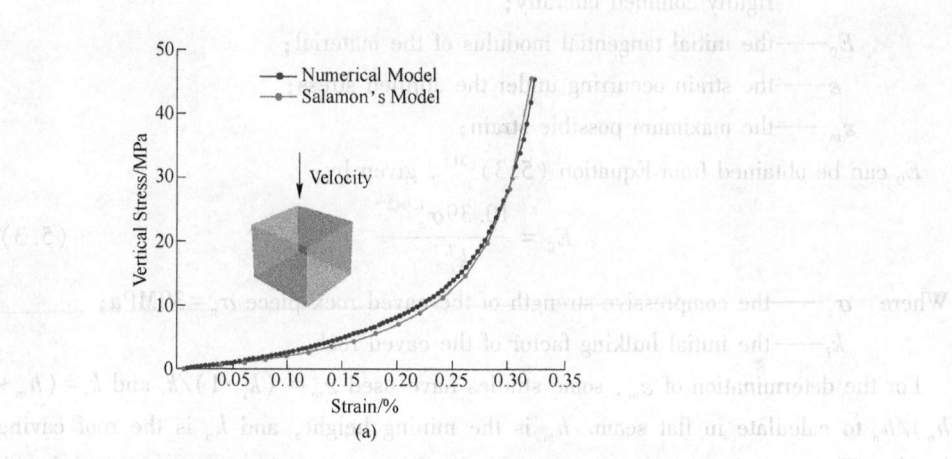

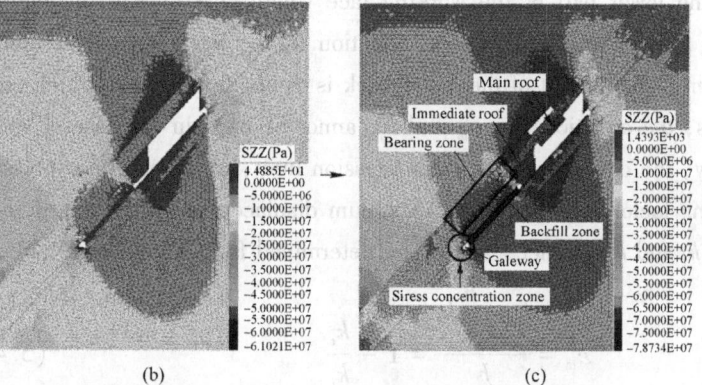 

The color pictures can be seen by scanning the QR code.

Figure 5.3　Stress redistribution process

(a) Stress-strain curve of the numerical model; (b) Stress-strain curve for Salamon's model for gob material; (c) Stress distribution

Table 5.1　Mechanical properties of the gob material

Material	Constitutive model	Properties						
		Cap pressure /MPa	ε_m	E_0 /MPa	Dens /kg · m^{-3}	Bu /GPa	Sh /GPa	Con /MPa	Fri /(°)
Gob filling material	Double-yield	$P=\dfrac{E_0\varepsilon}{1-\dfrac{\varepsilon}{\varepsilon_m}}$	0.36	15.8	1150	2	1	0.01	2

5.3.2.3 Stress Redistribution Results

The vertical stress distribution characteristics are shown in Figure 5.3(b) and (c). It can be seen that significant asymmetric vertical stress is exhibited in the roof and floor. In addition, the mining-induced stress of the overlying rock strata will be released and transferred after mining occurs, and its stress state will become simpler. Then, the filling caved rock will change the stress state to a three-dimensional stress state (stress restoration)[32] and it will exert more pressure. So, a large bearing zone forms above the backfill zone, as shown in Figure 5.3(c), and it will distribute more of the overburden pressure. However, the stress concentration zone is distributed around the gateway, as shown in Figure 5.2(c), and due to the role of the bearing zone, the degree of stress concentration degree is relative small (concentrated stress = $10 \sim 15$MPa, and gateway stress at the buried depth = 12MPa).

After the stress redistribution analysis, it was concluded that a low stress environment was formed around the gateway in a steep coal seam. This is advantageous for GER, and it is also useful in avoiding floor heave and coal bumps.

5.4 The Key Technology for GER

5.4.1 Support Zone for Steep Coal Seams

The maintenance of a gateway by GER is mainly conducted in two stages as follows: (1) entry support strengthening during the mining period; and (2) construction of a gob-side support zone, as shown in Figure 5.4.

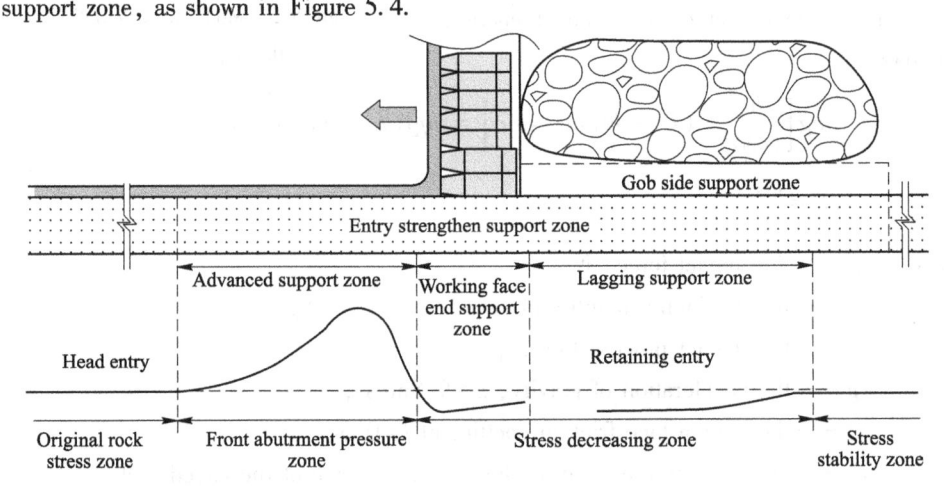

Figure 5.4   The gob-side entry retaining zone

The entry support strengthening zone includes the advanced support zone, the working face end support zone, and the lagging support zone. The former is mainly located in the front abutment pressure zone. The latter two are mainly located in the decreasing stress zone and seriously influenced by roof weighting and rotation, especially in the case of the working face end support zone. However, the influence is not that much greater than that of a flat and gently inclined coal seam, as analyzed above.

The gob-side support zone is located behind the end of the working face. Its filling material, which is significantly related to the dip of the coal seam, is usually high-water material (HWM), concrete, paste, block, and waste pack[5] when the dip angle is relative small. However, for a steep angle coal seam, so far, only natural filling with caved waste rock can be used to advantage.

5.4.2 Natural Filling Technology for Caved Waste Rock

The natural filling of caved waste rock is superior compared to other support methods. However, when rolling or sliding to the gob-side, as shown in Figure 5.5, the waste rock will have a high velocity and strong impact force. Here, taking into account the impact effect of the caved rock, the impact force was calculated to facilitate the difficult analysis of the current support and the later theoretical analysis of the supporting devices.

The impact force is calculated by Equation (5.5) according to the law of conservation of energy, as Follows:

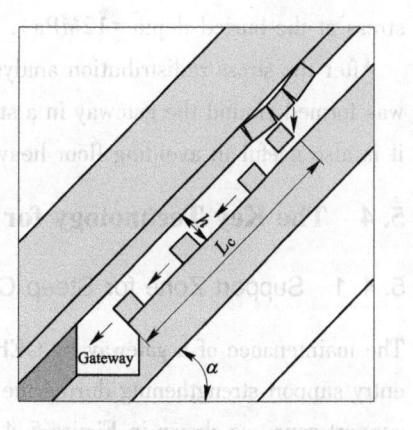

Figure 5.5 Model for calculating the impact force

$$\begin{cases} mg\left[\dfrac{h}{\cos\alpha} + (L_c - h\tan\alpha)\sin\alpha\right] = mgf(L_c - h\tan\alpha)\cos\alpha + \dfrac{mv^2}{2} \\ F = \dfrac{mv}{t} \end{cases} \tag{5.5}$$

Where F——the impact force, N;

m——the maximum quantity of caved waste rock, kg;

t——the impact process time, s;

g——the acceleration of gravity, $g = 9.8 \text{m/s}^2$;

f——the bottom face friction coefficient $f = 0.5$;

L_c——approximately treated as the rolling distance of the caved rock.

Here, to calculate the impact force, we utilized the following steep coal seam parameters (Deng and Wang, 2014): $L = 86m$, $\bar{h} = 2m$, $\alpha = 47.2°$, $h = 2.36m$, $m = (2 \times 4 \times 1.5)$ m³ × 2500kg/m³ = 30000kg, and $t = 2s$ (Zhou et al., 2012). According to the results of Equations (5.1) and (5.3), the impact force (F) of the largest caved waste rock against the support system is estimated to be 280kN. The minimum impact width of the waste rock is 1.5m, so the impact force over one meter is about 186.7kN. The calculation results indicate that only single hydraulic props temporarily support the GER is difficult to prevent the impact of waste rocks. Measures must be taken to reduce the impact force by increasing the impact time[21].

Thus, a flexible support technique[2] was used in the field, as shown in Figure 5.6.

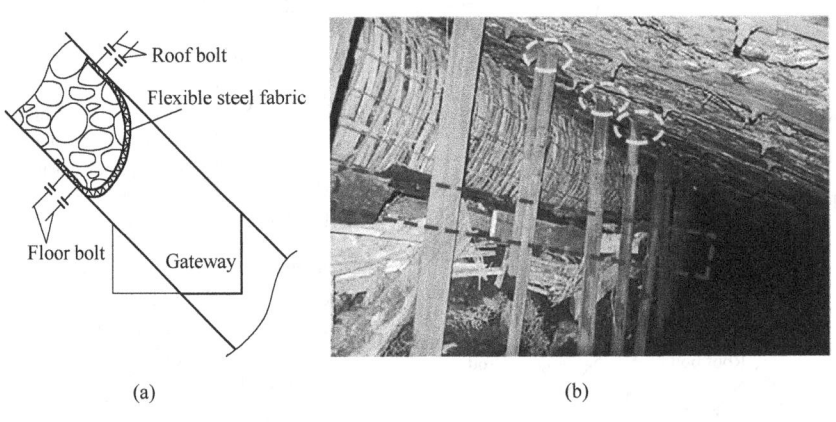

(a)   (b)

Figure 5.6   Flexible support technique
(a) Flexible steel fabric block; (b) Field effect

In practice, as shown in Figure 5.6(a), first bolts were set in the roof and floor, and then, the two ends of the flexible steel fabric were fixed to these bolts. However, there was large amount of weak interlayer or broken material due to the geological conditions of the floor, which can easily lead to the floor sliding along the bedding plane. This would make the floor bolt useless for fixing the flexible steel fabric in place, as indicated by the red zone in Figure 5.6(b). In addition, the single props are usually inserted into the roof, as indicated by the yellow zone in Figure 5.6(b)[2]. A new rock blocking device and grouting reinforcement method were invented to solve this problem, as shown in Figure 5.7.

As seen in Figure 5.8, the equipment can be divided into the gateway and gob-side rock blocking devices. The force of the roof's antiskid support plate (RASP) and the floor's antiskid support plate (FASP) are shown Figure 5.7(a). Their equilibrium limit relationships are shown as follows:

· 74 ·　5　Natural Filling and Systematic Roof Control Technology for Retaining Gateway in Steep Coal Seams

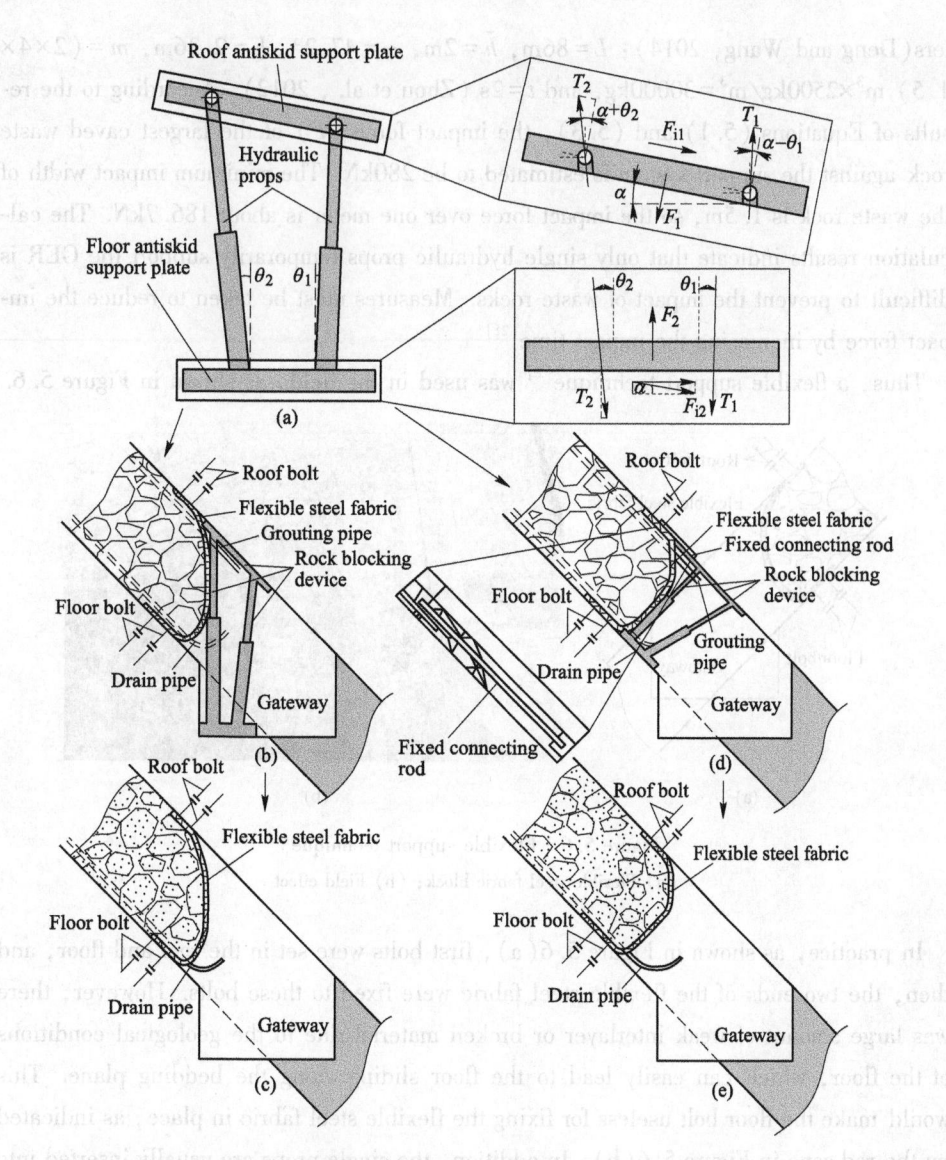

Figure 5.7　Rock blocking device and grouting reinforcement method

(a) Rock blocking device; (b) Gateway blocking device;

(c) Grouting effect; (d) Gob-side blocking device; (e) Grouting effect

$$F_1 w + T_1 \sin(\alpha - \theta_1) + T_2 \sin(\alpha + \theta_2) = F_{i1} \qquad (5.6)$$

$$F_2 w + T_1 \sin\theta_1 = F_{i2} \cos\alpha + T_2 \sin\theta_2 \qquad (5.7)$$

Where F_1 is the roof pressure, and F_2 is the floor pressure, as expressed by:

$$F_1 = T_2 \cos(\alpha + \theta_2) + T_1 \cos(\alpha + \theta_1) \qquad (5.8)$$

$$F_2 = T_2 \cos\theta_2 + T_1 \cos\theta_1 + F_{i2} \sin\alpha \qquad (5.9)$$

Where T_1, T_2——the pillar support force;

w——the friction coefficient between the RASP and the roof or between the FASP and the floor;

α——the dip of the coal seam;

θ_1, θ_2——the pillar tilt angles;

F_{i1}, F_{i2}——the impact forces caused by the caved waste rock.

It can be seen that the equilibrium relationship of the rock blocking device is complex and is related to several factors. When the rock blocking device is utilized in the gateway, as shown in Figure 5.7(b), and $\alpha > 0$, Equations (5.6) and (5.7) show that increasing T_2 will increase the stability of the RASP, but will decrease the stability of the FASP. However, when $\alpha > \theta_1$, increasing T_1 can improve the instability of the FASP and further increase the stability of the RASP. Thus, the two tilted hydraulic props hinged at the RASP and FASP can form a triangular stability support structure. Furthermore, Equation (5.10) can be obtained from Equations (5.6) to (5.9):

$$\frac{2w\left(T_2\cos\frac{\alpha+2\theta_2}{2}\cos\frac{\alpha}{2} + T_1\cos\frac{\alpha+2\theta_1}{2}\cos\frac{\alpha}{2}\right) + 2\left(T_1\sin\frac{\alpha}{2}\cos\frac{\alpha-2\theta_1}{2} + T_2\sin\frac{\alpha}{2}\cos\frac{\alpha+2\theta_1}{2}\right)}{F_{i1} + F_{i2}(\cos\alpha - \sin\alpha)} = n_1 \qquad (5.10)$$

In addition, when $\alpha = 0$, this device can be utilized at the gob-side, as shown in Figure 5.7(d). Thus, Equation (5.10) can be transformed into:

$$\frac{2w(T_2\cos\theta_2 + T_1\cos\theta_1)}{F_{i1} + F_{i2}} = n_2 \qquad (5.11)$$

Where n_1 and n_2 are the safety factors. Compared to Equations (5.10) and (5.11) when $\alpha = 47.2°$, $w = 0.3$ (Jin-an and Jun-ling, 2016), $\theta_1 = \theta_2 = 20°$, $T_1 = T_2 = 200\text{kN}$, and $F_{i1} + F_{i2} = F$, as seen in Equation (5.5), we found that:

$$n_1 > n_2 > 1 \qquad (5.12)$$

Thus, placing the device in the gateway can achieve greater stability than placing it at the gob-side. In reality, the RASP and FASP are for nonskid purposes and the friction coefficient (w) can be more than 0.3. Based on current technological capabilities, the value of T_1 and T_2 can be greater than 200kN; and combined with the stop function of the flexible steel fabric, the safety factor should be greater.

Furthermore, when used in steep coal seams, this method presents other features:

(1) The rock blocking device can be recycled after the grouting reinforcement, and when used at the gob-side, the initial stability can be well controlled by fixing the con-

· 76 · 5 Natural Filling and Systematic Roof Control Technology for Retaining Gateway in Steep Coal Seams

necting rod to the roof bolt, as shown in Figure 5. 7(d).

(2) The blocking device can satisfactorily support and maintain the stability and integrity of the retained gateway roof, especially for a soft roof, because of its integral structure and support capacity.

(3) The slurry can not only reinforce the broken waste rock and increase its bearing capacity to hinder deformation of the overlying rock strata[33], but it can also infiltrate into the fractured floor and improve the strength of the rock mass, allowing the blocking device to be fixed to the floor.

5.4.3 Strengthening Support Device

The stability and integrity of the roof's rock mass in strengthening support zone has a significant effect on the success of GER. So far, the vast majority of strengthening support zones have been supported by two or three rows of single hydraulic props with the articulated roof beam serving as the temporary support. However, because the props cannot form an integral structure, they will lack stability due to abutment pressure and roof rotation and weighting. In addition, a single hydraulic prop is usually inserted into the floor or roof because of the existence of uneven pressure due to the low strength and stiffness of the roof's rock mass and the higher strength and stiffness of the strengthening support body, which will negatively influence the quality of the support. Thus, the strengthening support zone in a steep coal seam requires a hydraulic support to provide increased strength to the roof's rock mass, as shown in Figure 5. 8.

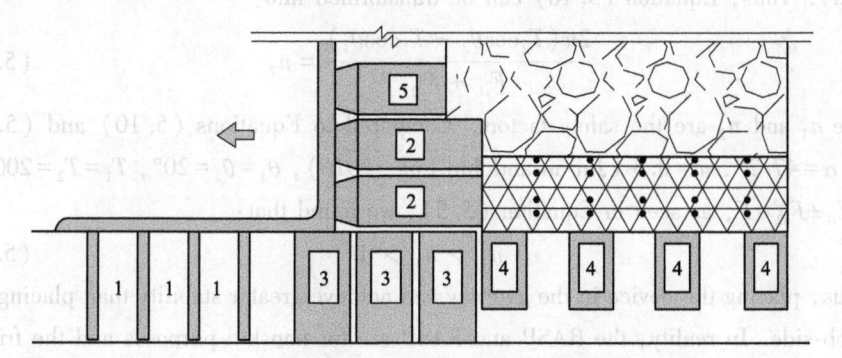

Figure 5. 8 Diagram of the entry strengthening support device

1—Advanced support hydraulic support;2, 3—Working face end hydraulic support;
4—Lagging hydraulic support;5—Ordinary hydraulic support

The advanced hydraulic support (1) has an integral structure, as shown in Figure 5. 9, so it can better resist mining-induced stress and large roof loading. Hence, it can improve the roof's integrity to benefit the GER.

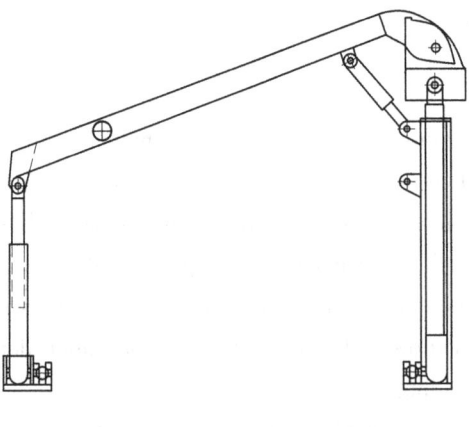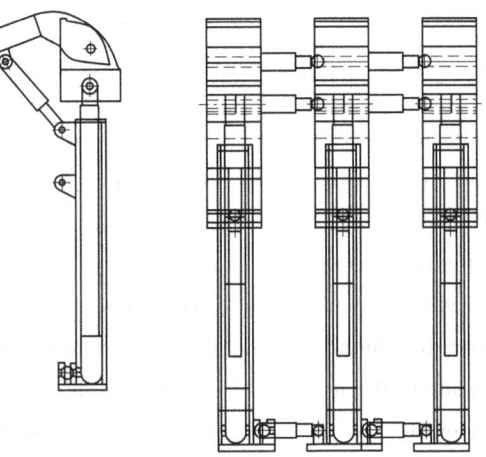

Figure 5.9   The advanced hydraulic support

The working face end support zone experienced serious roof deformation, but it also requires a large space with no pillar support for the retaining gateway. This makes it very difficult to maintain the roof at the end of the working face. To facilitate the implementation of the GER and later maintenance of the retained gateway, hydraulic supports (2) and (3) at the end of the working face were developed to fulfill this role. The details of their development are included in the patents (Numbers CN204327145U and CN202628152U, China, respectively).

The lagging support zone was affected by the roof cave-in. The lagging hydraulic support (4) has a simple and compact structural design, and it is easy to operate. It can provide significant supportive force and improve safety conditions, the details of which are included in the patent (Number CN203717003U, China).

5.5   Gateway Driving and Its Support Technology

The development of GER has been a systematic project, which not only involves support in the late retaining stage, but also includes the design of a cross section and driving support for a gateway in the initial driving stage[34]. These practices should be used together so as to ensure the stability and integrity of the roof in the initial stage and the success of the GER of the gateway in a steep coal seam. However, all of the retained gateways were implemented based on the current existing support conditions of gateways in China, which include expanding ribs and supplementary bolts (cable). This leads to the need for a large amount of material, is labor intensive, and expensive. Thus, the

· 78 ·   5 Natural Filling and Systematic Roof Control Technology for Retaining Gateway in Steep Coal Seams

size and shape of the gateway's cross section should be considered when choosing a support method.

5.5.1 Gateway Driving

Comprehensive mechanized driving is mainly accomplished by mechanical cutting and vibration. The degree to which this method disturbs the surrounding rock is far less than that of blast driving, so it causes less damage to the coal and rock masses. Thus, the comprehensive mechanized driving method is recommended for gateway driving. If the blasting method must be used, the borehole arrangement and blasting parameters should be chosen to minimize the number of fractures and to control their propagation depth so as to maintain the integrity of the surrounding rock.

Fully-mechanized longwall mining was successfully utilized in the Lvshui Dong mine in China with coal seam dips of 60°. As China gradually implements the development of fully-mechanized longwall mining and pillarless mining, more requirements have been put in for the size and shape of the gateway's cross section. In this study, we conclude that the the size (bottom width and roof height) of the cross section of a new gateway would not influence the production or the ability to meet any related requirements. In addition, field studies conducted by engineers on 75 statistical gateways (shown in Table 5.2) in Sichuan Province, China show that a trapezoidal cross section, as shown in Figure 5.10(a), is preferred when $\alpha \leqslant 40°$, and a special cross section, as shown in Figure 5.10(b), and an inclined arch cross section, as shown in Figure 5.10(c), are preferred when $\alpha > 40°$.

Table 5.2   Investigation of applications of the gateway cross section

Cross section	Trapezoidal cross section	Special cross section	Inclined arch cross section	Other cross section
Number	38	4	22	11
Proportion/%	50.7	5.3	29.3	14.7
Dip angle	≤40°	>40°	>40°	Others

In reality, the height of high side wall of a trapezoidal cross section gateway will increase as the dip angle increases, resulting in a large free face and simple stress state, which are prone to experience tensile and shear failure. However, for a gateway with a special cross section or an inclined arch cross section, the coal and rock masses will have a small free face and will experience multiaxial stress, which is conducive to increased stability. Thus, based on the current support technique, we chose a trapezoidal cross section when the dip of the coal seam is $35° < \alpha \leqslant 45°$, a special cross section when the surrounding rock is stable, and an inclined arch cross section when the surrounding rock is unstable and $45° < \alpha \leqslant 55°$, as shown in Figure 5. 10.

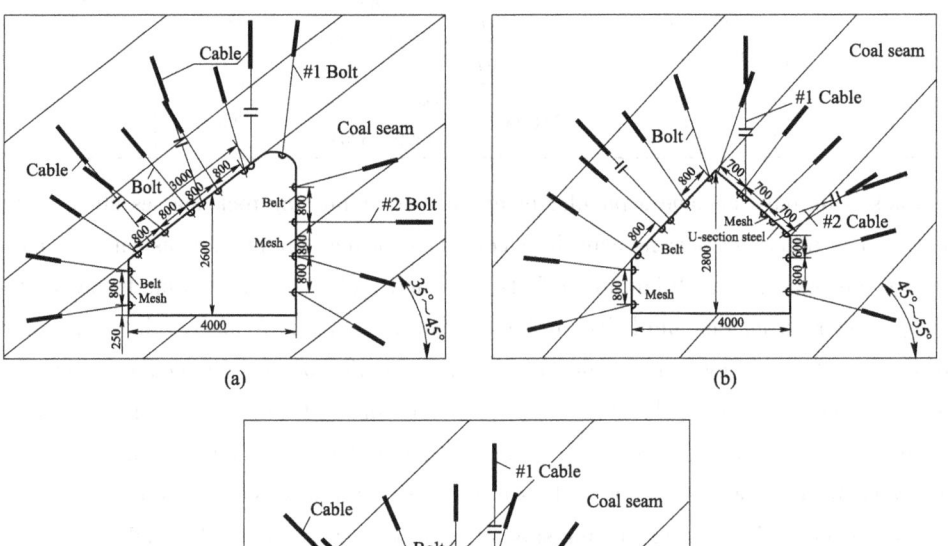

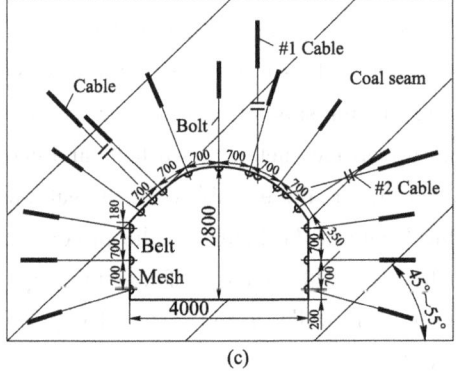

Figure 5. 10   Initial driving support system

5. 5. 2 Gateway Support Technology

Based on the above analysis, the supporting factors favoring the use of GER in steep coal seams can be summarized as follows:

(1) It ensures reliable support during the initial driving stage, which reduces the deformation and degree of fracturing of the surrounding rock.

(2) It to minimize the need for supplementary bolts (cable) to strengthen the support zone of GER, making efficient use of the existing bolts and cables.

Thus, the gateway support system is provided in the initial driving stage, as shown in Figure 5. 10. The detailed support parameters are listed in Table 5. 3.

5 Natural Filling and Systematic Roof Control Technology for Retaining Gateway in Steep Coal Seams

Table 5.3 Detailed support parameters

Initial driving support system	Material	Specification /mm	Inter-row spacing/mm×mm	Pretension force/kN	Anchorage length/m
Bolt in roof	High-strength left-screw- thread steel	$\varphi 22 \times (\geqslant 2400)$	800×800 or 700×700	90	1.65
Bolt in high side wall		$\varphi 22 \times (\geqslant 2400)$		60	1.65
Bolt in low side wall		$\varphi 22 \times (\geqslant 2000)$		60	1.43
Cable in roof	Steel wire	$\varphi 22 \times 6300$	(1500-2500)× (1500-2500)	200	2

Rock bolts have become a popular technique for reinforcing rock masses all over the world. The important bearing capacity mechanisms of rock bolting are suspending, nailing, beam building, arch building effects, and new rock mass theory (the behavior of a complex of rock masses and bolts can be assumed as a new rock mass with an increased RMR value)[35]. However, according to some statistics on the retained gateway, the bolt material, anchorage length, and pretension force of the initial driving support system usually result in poor stability and greater deformation during retaining of the gateway. In addition, one or two cables or bolts within a cross section were usually supplemented beyond the strengthening support zone to increase the roof's stability.

Thus, we improved the support parameters, used bolts and cables combined with steel mesh and steel belts, designed three cables in a gateway cross section, and appropriately increased the length of the bolts and the pretension force of the bolts and cables. In addition, the broken width of the roof and the upper side wall are usually greater than that of the lower side wall, so the roof and upper side wall require longer bolts than the lower side wall.

The support system can form a combined arch structure in the surrounding coal and rock[26,35], and the support parameters can enhance the bearing capacity of the combined arch structure. In addition, it uses an extensive bolting method, and the anchorage length is greater than 65% of the bolt length. This can increase the pre-stress sphere. The effect of the pre-stress anchor was more pronounced in the soft rock than in hard rock[36]. Cable reinforcement must be added to the support system to produce an o-verhanging effect in the stable roof rocks. In summary, the support system can improve the stress state of the surrounding coal and rock masses, and increase the thickness of the reinforcement structure, especially the arch foot, formed by the anchoring force, which will increase its self-bearing ability[37], and avoid large displacement and/or rapid failure.

Field studies have shown that the developed bolt materials and manufacturing processes, the high pretension force, and the intensive bolts and cables can effectively control the dilatant deformation and maintain stability.

In addition, the #1 and #2 bolts when appropriately extended and the #1 and #2 cables can be reserved and reused to fix the flexible steel fabric after the work face is advanced, so that the roof bolt and floor bolt (shown in Figure 5.7) would not be arranged for retaining the gateway.

5.6 Chapter Summary

This paper analyzed the support technology for GER in steep coal seams based on a systematic project. In detail, roof cave-in structures, filling characteristics, dynamic effects, and the roof's stress redistribution characteristics indicate that the use of GER in steep coal seams provides significant advantages. In addition, for natural filling, the rock blocking device and grouting reinforcement method developed in this study hold up well against the impact force of the waste rock and reinforce the waste rock and soft floor. Furthermore, a set of strengthening support devices were developed for use in the strengthening support zone. In addition, to more effectively retain the gateway in the late retaining stage, the size (bottom width and roof height) of the cross section of a new gateway does not influence the production or the ability to meet any related requirements. The appropriate type of cross section, i. e. , Trapezoidal cross section, special cross section, or inclined arch cross section, was determined according to the dip of the coal and the type of surrounding rock. Finally, based on two driving support principles, a support system consisting of bolts and cables combined with steel mesh and steel belts was developed. The optimized bolt material, increased length of the bolts and their anchorage length, and the increased pretension force of the bolts and cables were the main features of the support system that improved the self-bearing ability of the surrounding coal and rock masses.

References

[1] Yang H, Cao S, Wang S, et al. Adaptation assessment of gob-side entry retaining based on geological factors[J]. Eng Geol, 2016, 209: 143-151.

[2] Zhang Y, Tang J, Shen P, et al. Flexible support technique for the weak roof and floor of a roadway in a coal seam with a medium-thickness and a large inclined angle[J]. Mode Tunnel Technol, 2015, 52:198-204.

[3] Cao S G, Zou D J, Bai Y J, et al. Study on upward mining of sublevels for gob-side entry retaining in three-soft thin coal seam group[J]. J Min Saf Eng, 2012, 29: 322-327.

[4] Li X, Ju M, Yao Q, et al. Numerical investigation of the effect of the location of critical rock

block fracture on crack evolution in a gob-side filling wall[J]. Rock Mech Rock Eng, 2016, 49: 1041-1058.

[5] Tan Y L, Yu F H, Ning J G, et al. Design and construction of entry retaining wall along a gob side under hard roof stratum[J]. Int J Rock Mech Min Sci, 2015, 77:115-121.

[6] Yang J, Cao S, Li X. Failure laws of narrow pillar and asymmetric control technique of gob-side entry driving in island coal face[J]. Int J Min Sci Technol, 2013, 23: 267-272.

[7] Zhang J X, Miao X X, Guo G L. Development status of backfilling technology using raw waste in coal mining[J]. J Min Safety Eng, 2009, 26:395-401.

[8] Hu B N, Guo A G. Testing study on coal waste back filling material compression simulation[J]. J China Coal Soc, 2009, 34:1076-1080.

[9] Jiang Z Q, Ji L J, Zuo R S. Research on mechanism of crushing-compression of coal waste[J]. J China U Min Technol, 2001, 30:139-142.

[10] Miao X X, Zhang J X. Analysis of strata behavior in the process of coal mining by gangue back-filling[J]. J Min Safety Eng, 2007, 24:379-382.

[11] Su Q Z, Hao H J. Research on adaptability of deformation characteristics of roof and packing compressibility[J]. J Jiaozuo Ins Technol, 2002, 21:321-323.

[12] Cao S G. The analysis of mechanical construction of surrounding rocks in the pitching Seam[J]. J Chongqing U, 1992, 15:128-133.

[13] Huang J G. Structural Analysis for roof movement for steep coal seams[J]. J China U Min Technol, 2002, 31:411-414.

[14] Wu Y P, Xie P S, Ren S G. Analysis of asymmetric structure around coal face of steeply dipping seam mining[J]. J China Coal Soc, 2010, 35:182-184.

[15] Wu Y P, Xie P S, Wang H W, et al. Incline masonry structure around the coal face of steeply dipping seam mining[J]. J China Coal Soc, 2010, 35:1252-1256.

[16] Gou P F, Xin Y J. Stability analysis of roof structure in pitched seam gateway[J]. J China Coal Soc, 2011, 36:1607-1611.

[17] Huang Q X, Dong B L, Chen G H, et al. Failure mechanism of entry in steep soft seam and bolting design[J]. J Min Safety Eng, 2006, 23:333-336.

[18] Xin Y J, Gou P F, Yun D F, et al. Instability characteristics and support analysis on surrounding rock of soft rock gateway in high-pitched seam[J]. J Min Safety Eng, 2012, 29:637-643.

[19] Deng Y, Wang S. Feasibility analysis of gob-side entry retaining on a working face in a steep coal seam[J]. Int J Min Sci Technol, 2014, 24:499-503.

[20] Zhang Y, Tang J, Xiao D, et al. Spontaneous caving and gob-side entry retaining of thin seam with large inclined angle. Int J Min Sci Technol, 2014, 24: 441-445.

[21] Zhou B, Xu J, Zhao M, et al. Stability study on naturally filling body in gob-side entry retaining [J]. Int J Min Sci Technol, 2012, 22: 423-427.

[22] Itasca. 3DEC——3 dimensional distinct element code [D]. Minneapolis: Itasca Consulting Group Inc, 2013.

[23] Wu Y, Yun D, Zhang M. Study on the elementary problems of full-mechanized coal mining in

greater pitching seam[J]. J China Coal Soc, 2000, 25:465-468.

[24] Yang H, Cao S, Li Y, et al. Assessment of excavation broken zone around gateways under various geological conditions: a case study in Sichuan Province, China[J]. Minerals, 2016, 6:72.

[25] Wu Y P, Liu K Z, Yun D F, et al. Research progress on the safe and efficient mining technology of steeply dipping seam[J]. J China Coal Soc, 2014, 39:1611-1618.

[26] Tu H, Tu S, Yuan Y, et al. Present situation of fully mechanized mining technology for steeply inclined coal seams in China[J]. Arab J Geoscl, 2015, 8:4485-4494.

[27] Gao F, Stead D, Kang H, et al. Discrete element modelling of deformation and damage of a roadway driven along an unstable goaf——a case study [J]. Int J Coal Geol, 2014, 127: 100-110.

[28] Shabanimashcool M, Li C C. Numerical modelling of longwall mining and stability analysis of the gates in a coal mine[J]. Int J Rock Mech Min Sci, 2012, 51:24-34.

[29] Shabanimashcool M, Li C C. A numerical study of stress changes in barrier pillars and a border area in a longwall coal mine[J]. Int J Coal Geol, 2013, 106:39-47.

[30] Wang H, Jiang Y, Xue S, et al. Assessment of excavation damaged zone around roadways under dynamic pressure induced by an active mining process[J]. Int J Rock Mech Min Sci, 2015, 77: 265-277.

[31] Yavuz H. An estimation method for cover pressure re-establishment distance and pressure distribution in the goaf of longwall coal mines[J]. Int J Rock Mech Min Sci, 2004, 41:193-205.

[32] Li Y M, Liu C Y, Li X M, et al. Roof control effect for gangue backfilling of goaf in thin steeply inclined seam under water body[J]. J China Coal Soc, 2010, 35:1419-1424.

[33] Li X S, Xu J L, Zhu W B, et al. Simulation of backfill compaction character by particle flow code[J]. J China Coal Soc, 2008, 33:373-377.

[34] Hua X Z. Development status and improved proposals on gob-side entry retaining support technology in China[J]. Coal Sci Technol, 2006, 34:78-81.

[35] Mohammadi M, Hossaini M F, Bagloo H. Rock bolt supporting factor: rock bolting capability of rock mass[J]. B Eng Geol Environ, 2017, 76:231-239.

[36] Zheng X G, Zhang N, Xue F. Study on stress distribution law in anchoring section of prestressed bolt[J]. J Min Safety Eng, 2012, 29:365-370.

[37] Yang S S. Study on the surrounding rock control theory of roadway in coal mine[J]. Journal of the China Coal Society, 2010, 35(11):1842-1853.

6　Soft Roof Failure Mechanism and Supporting Method for Retaining Gateway

6.1　Introduction

Previous studies showed that the gateway roof rock mass failure mechanism and supporting methods had a significant impact on gob-side entry retaining, determining the procedure's success. Many researchers categorized five failure types of a gateway roof rock mass, including the compressive stress failure type, tensile stress failure type, shear stress failure type, squeezing and fluidity failure type, and geological structure failure type[1-3]. Based on the above failure mechanism of gateway roof strata, many active support models were suggested, and a combined supporting technology using a bolt, mesh, and cable was found to be effective for ground control. It was found that a combined support was unable to control the main roof in roadway retaining according to the theory of 'given deformation', but it can stabilize the immediate roof with the main roof to maintain the retained gob-side entry very well[4,5].

Rock bolts have become a popular technique for reinforcing rock masses all over the world. Rock bolts are installed to reinforce a fractured rock mass by resisting dilation or shear movement along the fractures. Nemcik et al. [6] determined that the non-linear bond-slip relationship used in the FLAC model for bolting accurately matches the experimental data reported by Ma et al. [7]. Yang[8] divided the anchored force evolution process into three stages and found that a surface structure-like bearing plate produced a maximum anchorage force acting directly on the surface of rock that could provide the greatest degree of support. This not only improved the rock mass stress state but also increased the thickness of the reinforcement structure formed by the anchoring force. Zheng and Zhang[9] determined the shear stress and pressure stress distribution equation for the anchored segment and upon studying the stress distribution rule of anchored segment, found that the effect of the pre-tensioned bolt in the soft rock was superior to that in hard rock. Cao[10] showed that the integrity of the supporting system that prevented local failure of surrounding rock from progressing into overall failure was important in rock bolting, so that reinforcing measures should be taken when necessary. Fan[11] designed a roadway heterogeneity controlling technology on a siltstone roof. Hua[12] made use of bolt support and anchor cable reinforcement support technologies inside and beside a retained roadway, respectively, and maintained the retained gob-side entry very well. Yan[13] used pre-tensioned bolts on a roadway roof with medium and fine-grained sandstones by an extrusion lockset to limit its vertical deformation for entry integrity and also analyzed the mechanical mechanism, technical principle and technical characteristics of bolt and cable coupling supports at a mine site. Chen and Bai[14] used bolts with high pre-tensioning, high strength, and a large elongation rate and cables as the basic supporting element, a single hydraulic prop with a metal hinge top beam as a reinforcement support inside the roadway, and a high-water rapid-solidifying material to support the side of the roadway to effectively control the roof rock deformation with medium and fine-grained sandstones in gob-side entry retaining.

In the past experimental studies and supporting theory[1,11-14], the bolt support is mainly aimed at the control of the medium-hard roof rock mass, which are generally medium and fine-grained sandstones in the horizontal and flat seams. The supporting theory also provides supporting references for the similar conditions of the roadway in the initial excavation stage and retaining stage. However, the working face end roof and retained roadway roof seriously deformed in the soft roof rock mass and gentle inclined coal seam with a soft roof rock after referencing the existed supporting theory.

As the second major geological factor that affects the adaptability of retaining gateway, roof integrity largely determines the effect of retaining gateway. For the gateyway roof with poor integrity, it is of great representative significance to study the roof failure behavior during retaining gateway. The integrity of the roof in a coal mine was 55 points. According to the score standard of the integrity index of the roof, the integrity of this roof was the worst. It had dense fracture distribution, spacing less than 0.4m, greater than 0.1m, and had low rock mass strength, which was a soft roof.

Therefore, to further develop our understanding of the soft roof failure mechanism and provide a supporting method, the following works were conducted in this chapter. First, the roof failure phenomenon of a gently inclined coal seam working face end and retained roadway were investigated at a coal mine in the southwest of China. Second, the roof failure mechanism was analyzed through the field investigation and stress evolution law obtained by three-dimensional distinct element code (3DEC). Finally, a roof support was designed based on results of this investigation to maintain roadway stability.

6.2 Description of Field Observation

6.2.1 Survey of Study Site

The target coal mine is located in the southwest of China. The mining coal seam is #C19, with an average dip angle of 16° and a mining thickness of 1.1m. The mining method is the fully mechanized longwall, and gob side entry retaining is used, as shown in Figure 6.1(a). The investigated head entry is buried at 315m, with a cross-section of 3.8m in width and 2.1m in height of the short rib, where resin-anchored rock bolts (ϕ20mm×2200mm), cables (ϕ12.54mm×6300mm), and reinforcement mesh were used, and three rows of single hydraulic supports with top hinged beams were used as advanced strengthening support, as shown in Figure 6.1(b). The roof rock bolts are non-fully anchored (anchor length is 1.4m) with a pre-tension of 90kN, inter-row spacing of 800mm×800mm and bolt angles of 90°, 80°, 70°, and 60° for different bolts, and the extra cables have a pre-tension of 200kN and an inter-row spacing of 3200mm×1200mm.

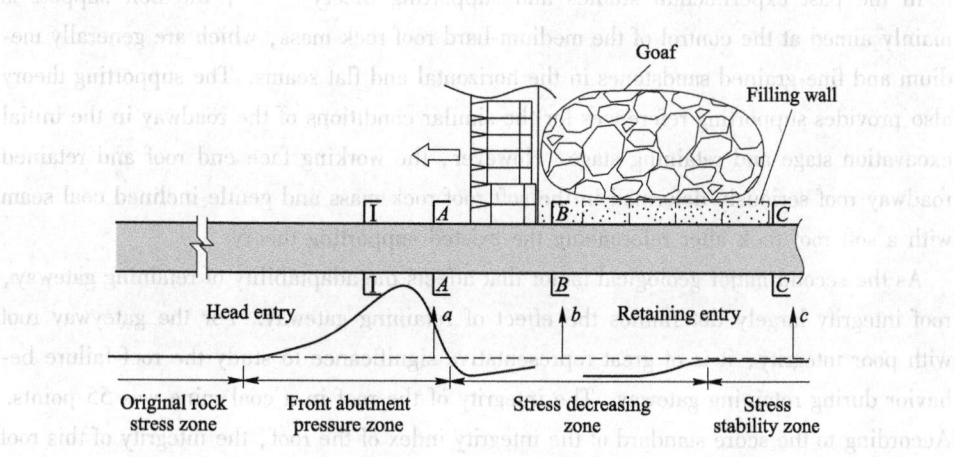

(a)

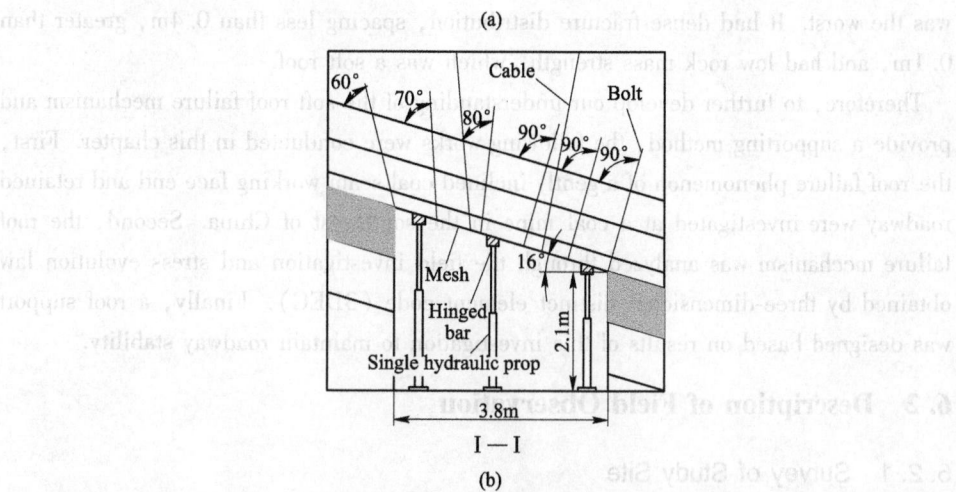

(b)

Figure 6.1 Schematic

(a) Gob-Side entry retaining; (b) Advanced strengthening support

According to a geological survey, the gateway roof strata typically consist of weak rocks. The immediate roof is approximately 1.0m in thickness and is made up of sandy mudstone and mudstone. The main roof consists of more competent siltstones and silty mudstones containing siderite nodules. In other words, the roof strata have low strength and poor stability and exhibit clear stratification, thus resulting in an immediate roof caving with the working face advancing as well as part of the main roof. Also, the floor is made up of sandy mudstone. It must be stressed that, at the working face end, the gateway soft roof was often affected by dynamic loading for the following reasons: (1) caving of the overburden rock mass in the gob roof, and the gangue movement, collision; (2) periodic weighting of the main roof; and (3) blasting operations of adjacent mining and excavating faces.

6.2.2 The Survey Results of Soft Roof Failure Characteristics

The soft roof failure characteristics of the front abutment pressure zone, the stress decreasing zone, and the stress stability zone are shown in Figure 6.2. According to a filed investigation of the head entry, there was a serious roof deformation due to the impact of front abutment pressure and a few cracks with less opening over a small area of the roof surface, as shown in Figure 6.2(a). With the working face advancing, the roof overhanging length increased, and cracks formed and joined together in the roof rock mass, as shown in Figure 6.2(b), resulting in a decrease of the roof strength, loss of self-bearing capacity, separation from the main roof, and roof deformation and becoming unload rock body, especially for the rock mass indicated by the red line. In addition, two bolts and one cable (B_1, B_2 and C_1) on the gob side lost their support ability. After

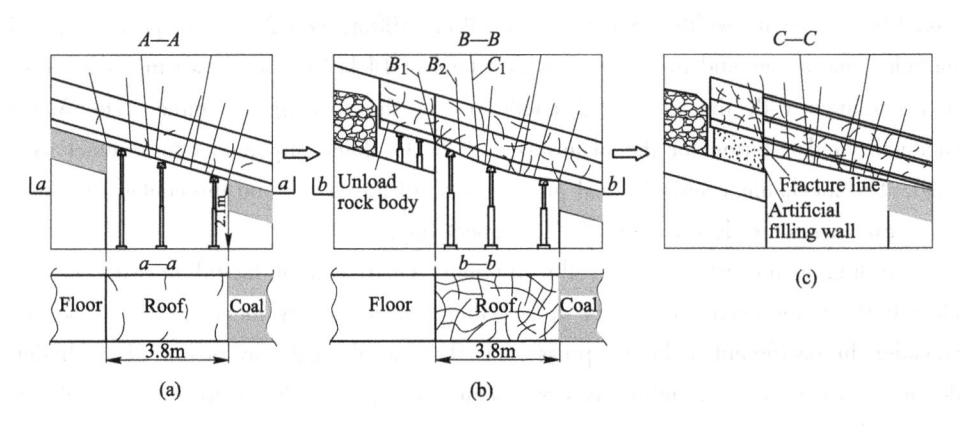

Figure 6.2 Schematic of soft roof rock mass failure

(a) In front abutment pressure zone; (b) In working face end; (c) In stress stability zone of retaining gateway

building the filling wall and removing the supports in the stress stability zone, the roof fractured along the inside of the filling wall, indicated as the 'fracture line' in Figure 6.2(c).

Field observations found that there were a large number of cracks in the vertical direction of roof near the end of the working face, so the roof caving occurred generally within a height range of 1.5m to 2m, as shown inFigure 6.3.

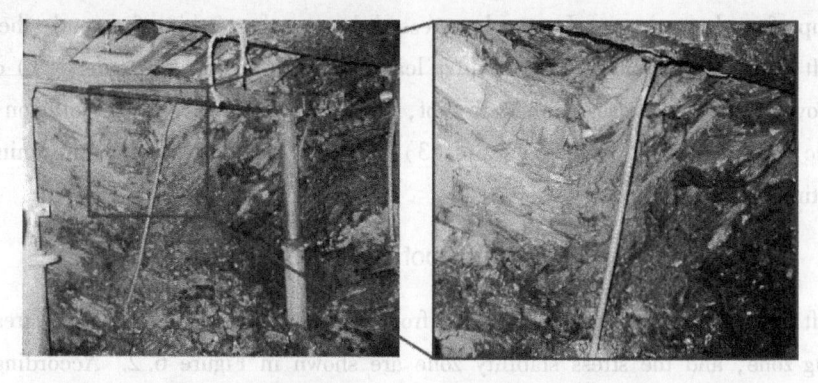

Figure 6.3   Caving roof near the working face end

6.3   Stress Evolution Law in Roof

To analyze the stress evolution law in the roof, the numerical modeling was adopted.

6.3.1   Numerical Simulation Model

Because of the coal seam condition and as the rock mass was discontinuous[15,16], 3DEC of ITASCA[17,18] was used to study the stress evolution law in the roof. The hexahedral model has a length, width, and height of 230m, 100m, and 200m, respectively, and includes coal seams and rock strata, with a total of 14 layers, as shown in Figure 6.4, in accordance with the geological conditions of the investigated mine. The Mohr-Coulomb yield criterion for the materials and the Coulomb slip model for contact were used. The mechanical and physical properties of all the layers and the contacts between every two layers are described in[10,19], respectively.

According to the reference[20], the minimum coefficient of lateral pressure is very close to 0 and the maximum can be up to 6 due to tectonic movement. For this case, we consider the coefficient of lateral pressure is 0.5, as the gateway is placed at shallow depth and close to the anticline axis and seam outcrop, which resulting in a much low horizontal stress. So, the state of the *in situ* stresses is $\sigma_x = \sigma_y = 4.25$MPa and $\sigma_z = 8.5$ MPa, with σ_y parallel to the longwall advance direction and σ_x perpendicular, as shown in Figure 6.4(b). A vertical pressure of 8.5 MPa is applied on the top surface, and the velocity of the bottom surface was restricted in all three directions, $v_x = v_y = v_z = 0$m/s. The velocity of the other four surfaces were restricted in the normal direction $v_n = 0$m/s.

The shape and size of the head entry are introduced in Section 6.2. After an initial equilibrium calculation, rock bolts wereinstalled, as shown in Figure 6.1(b), once the head was entry excavated. The rock bolts were represented as built-in 'cable' elements. For resin-grouted rock bolts, the stiffness (K_{bond}) and the cohesive strength (S_{bond}) of the grout are the two key properties that govern the anchor characteristics[19]; $K_{bond} = 3.06 \times 10^9$ N/m/m and $S_{bond} = 2.3 \times 10^5$ kN/m were adopted in this study.

In addition, a cross-sectional area of 3.142×10^{-4} m^2, an elastic modulus of 200GPa, and a tensile yield strength of 165kN were assigned to the 'cable' element. A stepwise excavation in the y-direction was adopted to simulate the working face advancing by deleting blocks in five steps of 10m each, as shown in Figure 6.4(c).

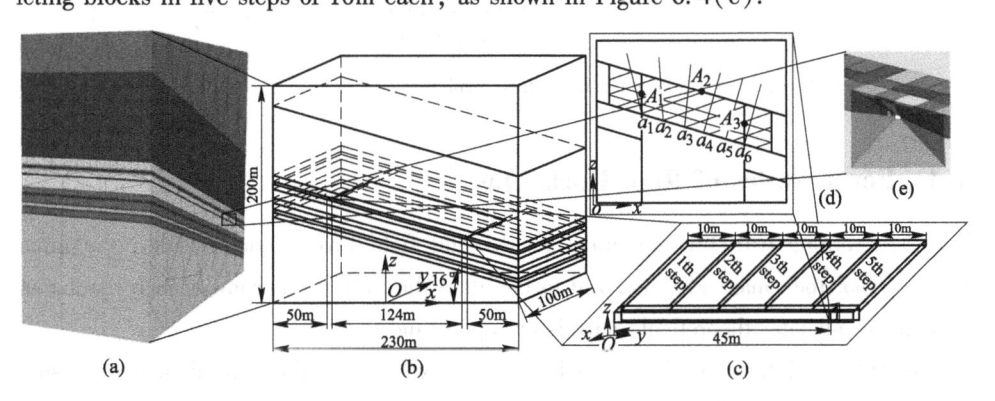

Figure 6.4   Calculation model
(a) Numerical model; (b) Coal seam mining model; (c) Working face advancing;
(d) Immediate roof grid, bolt and monitoring points; (e) Gateway cross section of model

6.3.2   Numerical Results

To understand the stress evolution law, the horizontal stresses in the x-direction of points A_1 and A_3 and the vertical stress of point A_2 in the roof at $y = 45$m, as shown in Figure 6.4(d), were monitored during the working face advancing. The results are shown in Figure 6.5. Especially, it can be found that from A to B Zone, the horizontal stress on the coal side (A_3) increases slightly in magnitude and then reduces to approximately 5.9MPa, while the horizontal stress on the gob side (A_1) drops instantaneously and appears tensile, and the vertical stress at the gateway central (A_2) reduces significantly to 4.5MPa as soon as the working face advances beyond the monitor points. The stresses in the gateway roof changed over the whole process from the beginning of the caving to the roof stabilization, until a tensile state was reached, which would influence the stability of the gob-side entry retaining.

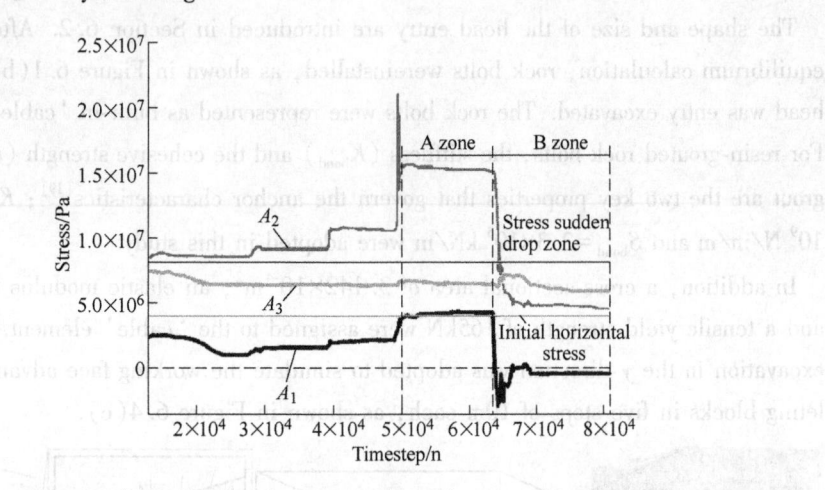

Figure 6.5　Roof rock mass stress evolution law in working face end

6.4　Force States of Roof Rock Mass

Here, combining with the numerical results and field investigation results, the force states of roof rock mass was analyzed to facilitate the failure mechanism analysis of roof rock mass and later theoretical analysis of supporting.

The stress evolution law showed in Figure 6.5 experienced the front abutment pressure zone (section A-A in Figure 6.1) and stress decreasing zone (section B-B in Figure 6.1). In section A-A, there are lateral forces F_{1a} and F_{1b} parallel to the rock strata and an overburden pressure P_1 resembling the stress (A_1, A_2 and A_3) in A Zone in Figure 6.5. In addition, field investigation showed that, there were bolt, mesh, and cable support force T_1, reinforced supporting force R_1 in advance, shear forces Q_{1a}, Q_{1b} on the two sides and gravity G. The force state is shown in Figure 6.6(a).

In section B-B, the gob side lateral force disappears along the strata strike direction resembling the stress (A_1) in B Zone in Figure 6.5, because of the roof rock mass caving near the gob side and part of the roof rock mass fracturing, but the coal side lateral force is F_{2b} resembling the stress (A_3) in B Zone in Figure 6.5. On the other hand, the overburden of pressure is assumed to become 0 resembling the stress (A_2) in B Zone in Figure 6.5, due to the roof separating from main roof. There are also a bolt, mesh, and cable support force T_2, a strengthen support force R_2 at the working face end and a shear force Q_{2b} in field, also the gravity G. The force state was shown in Figure 6.6(b).

Furthermore, the roof rock mass stress in cross section C-C, as shown in Figure 6.1, has experienced the stress states of sections A-A and B-B, but, field investigation showed that it would change after the construction of the artificial filling wall and the main roof rotary sinking, the lateral force F_a on the coal side and the overburden pressure P would restore, and the lateral force on the gob side would change to F_b. In addition, the bolt, mesh and cable support force becometo T, the shear forces at the two ends become to Q_{1a} and Q_{1b} and gravity remain the same G. The force state is shown in Figure 6.6(c).

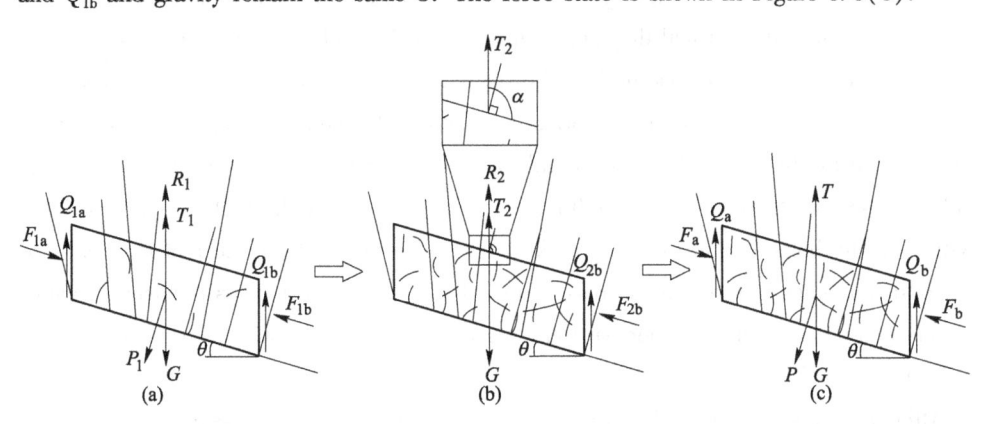

Figure 6.6   Roof rock mass force

(a) Force of cross section A-A; (b) Force of cross section B-B; (c) Force of cross section C-C

6.5   Mechanism for Failure of Roof Rock Mass

Roof failure was affected by many factors, so we would combine them with the failure characteristics obtained by filed investigation, stress evolution law obtained by numerical modeling and the force states to study the roof failure mechanism.

6.5.1   Mechanism for Failure of Roof Rock Mass in Working Face End

6.5.1.1   Unloading Effect of the Lateral Stress

Affected by the abutment pressure in the mining and excavation process, the roof rock mass is in the yield state, as shown in cross section A-A in Figure 6.1. The pre-tensioned bolt-cable-mesh supporting system improved the strength, significantly increased the yield strength, and altered the deformation characteristics of the roof rock mass. At the same time, the support system exerted a pressure stress on the rock mass, so the compressive zone stress state had to be altered, which could offset some of the tensile stress and friction and enhanced the shear capacity. In addition, the axial and lateral anchored forces increased the shear strength of the weak structural plane, preventing the roof rock mass from moving and sliding along the block structure plane. The pre-tensioned support system controlled the expansion deformation and destruction, preventing roof separation, sliding, fracture opening, and new crack generation in the anchorage zones, not only maintaining the integrity but also forming a pre-tensioned bearing structure with a large stiffness[21,22].

Influenced by the rotary sinking of the overlying roof and the superposition of the overburden pressure, the vertical displacement of the anchor bolt and cable increased significantly, leading to an increase of T_1. The stress increment of the roadway roof rock mass mainly comes from the bulking deformation. When the bolt, mesh, and cable apply pressure on the broken rock mass and add an anchoring force, the rock mass volume or volume rise rate decrease, and a bulking force is produced by the broken rock mass that will work the bolts and surrounding rock at the same time and put the roof rock mass in an extrusion state. As shown in cross section A-A in Figure 6.1, it is precisely because of the interaction of the 'support-surrounding rock', that the roof rock mass presented a failure state in Figure 6.2(a).

Although the single hydraulic props support the roof strata with pressure, the extrusion state rock mass above the hydraulic support started loosening on the working face end. In cross section B-B in Figure 6.1, the loosening roof rock mass caved behind the support body on the gob side, causing a sharp reduction of the lateral pressure, especially for the accumulated bulking force. Meanwhile, the overburden pressure decreased almost to 0 due to the separation of the soft roof from the hard roof above.

It can be observed that, accompanied by the gob roof caving, the lateral pressure unloaded and the bulking force decreased in the broken rock mass of the roadway roof, leading to its stress state changing from a three-dimensional stress state to a two-dimensional or uniaxial stress state, leading more easily to failure. At this point, together with the tensile stress effect, the rock mass volume 'elastic' expansion in the layer strike direction caused the generation and propagation of a large number of cracks and the further failure of the rock mass.

6.5.1.2 Tensile and Shear Failure of the Working Face End Lower Layer Rock Mass

Through the field investigation of the roof support body layout and its work process and based on the features of the roof rock mass deformation and failure, it is found that the

6.5 Mechanism for Failure of Roof Rock Mass

working face end soft roof presents the following phenomenon:

(1) An uneven supporting at the working face end roof. A strong mine ground pressure appeared, causing the significant sinking of the working face end roof in the working face advancing process, so a single hydraulic prop with an articulated roof beam was usually set as a reinforcement support. There is uneven pressure on the roof surface due to the low strength and stiffness of the sandy mudstone and mudstone and the higher strength and stiffness strengthening the support body, resulting in part of the support body inserting the roof and leading to a roof rock shear failure with shear stress q, as shown in Figure 6.7. In addition, the passive supporting force is too large and does not couple support with the bolts (cables), causing part of the bolts (cables) to be loosened, affecting the roof rock mass local stress state and resulting in ultimate rock mass failure.

(2) Effect of local moment. Treating the reinforcement body (individual hydraulic prop and articulated roof beam, *etc.*) as the fulcrum, the lower roof strata produced a local bending on a small scale. But when the fulcrum bending moment (M) is too large, the layered roof strata cause tensile and compressive damage to the upper surface and lower surface, respectively, additionally increased the possibility of the support body inserting the upper roof and leading to rock mass shear failure, as shown in Figure 6.7.

The two phenomena above are shown in the Figure 6.1 *A-A* and *B-B* cross sections and may occur simultaneously.

6.5.1.3 Dynamic Failure

The working face end roof rock mass is often affected by vibration as described above. Here, 3DEC was also applied to study the dynamic failure characteristics of the roof rock mass, the coal and rock mass physical and mechanical parameters, and the boundary conditions used in the numerical model shown in Figure 6.4, additionally adding the model viscous boundary conditions for all boundaries to make the stress wave propagate or be absorbed to simulate the infinite foundation environment. However, we alter the length to 1.6m in the y direction and treat it as a plane strain model. As the location in cross section *A-A* in Figure 6.1, under the action of a triangle stress wave to simulate the influence of a dynamic disturbance to the roof rock mass. In the simulation process, the stress peak value of the triangle stress wave is 8MPa rise time is 1ms and the decrease time is 7m/s[23-26]. Roof bolt, like a_1, a_2, a_3, a_4, a_5, a_6 in Figure 6.4(c), end displacement and the surrounding rock plastic zone were monitored, and the results

· 94 ·　6　Soft Roof Failure Mechanism and Supporting Method for Retaining Gateway

are shown in Figures 6.7~6.9, respectively.

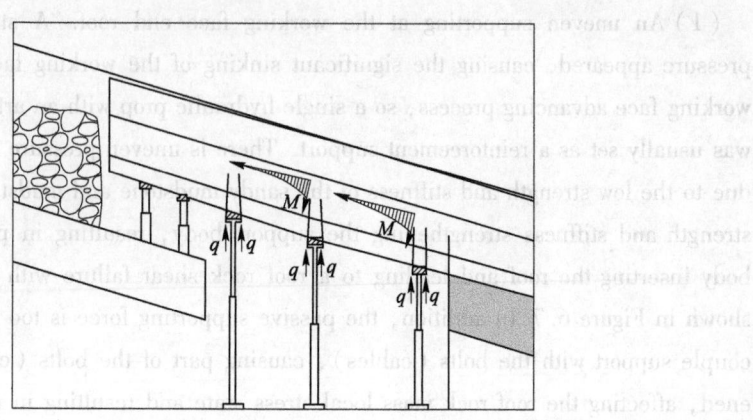

Figure 6.7　Schematic of support body damage to the roof rock mass

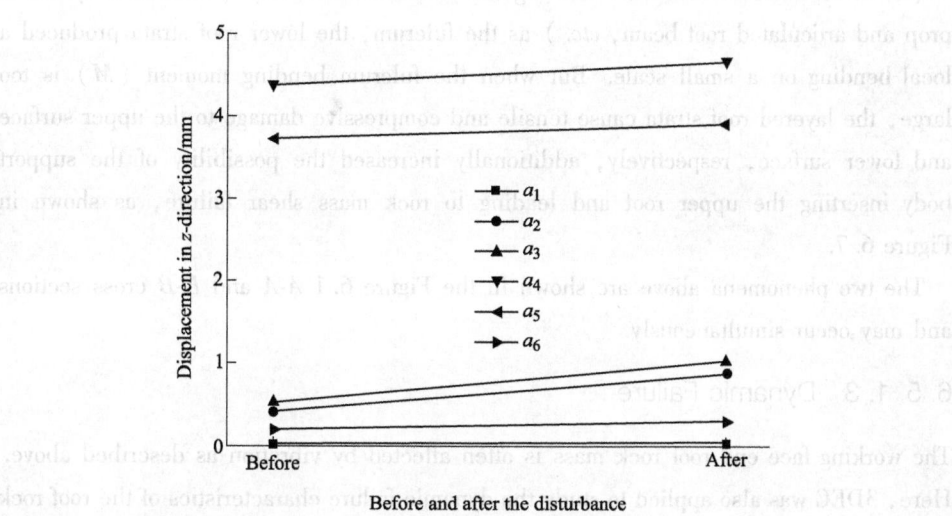

Figure 6.8　Displacement variation before and after the dynamic disturbance

It can be observed from Figure 6.8 that the displacement of all the monitoring points had a small increase in magnitude after the disturbance.

Figures 6.9(a) and (b) show the plastic distribution zone before and after the dynamic disturbance of the gateway surrounding rock mass, respectively. The stress wave had countless reflections and refractions on the roof crack surface according to the stress wave propagation theory, and the tensile stress would cause changes to the roof rock mass stress state and result in a combination of shear failure on existing joints/weakness horizons, an extension of critically oriented joints and propagation of new fractures

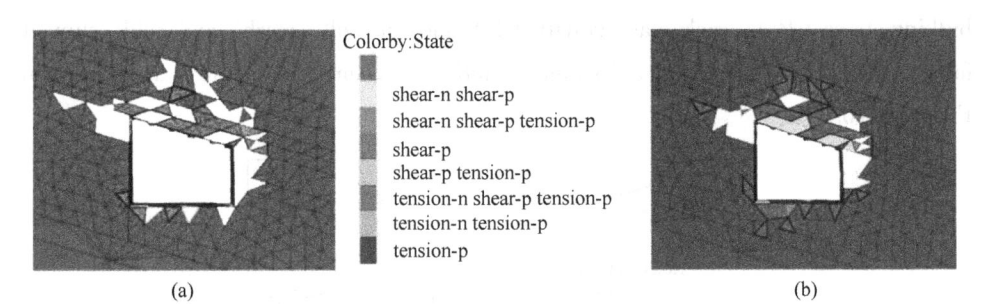

Figure 6.9　Plastic zone of surrounding rock

(a) Plastic zone before disturbance; (b) Plastic zone after disturbance

through previously intact rock so that the integrity of the roof became poor, the plastic zone expanding into the higher strata, and the ultimate widespread destruction of the roof rock mass[4]. At the same time, the two gateway sides and floor rock mass also produced plastic damage under the stress wave action, but had small extended failure zones.

The above results show that the increment of the displacement and plastic zone was small. But due to the constant increase of the dynamic load, it can be presumed that the working face end soft roof rock deformation and failure magnitude would significantly increase under the repeated dynamic load action, and the more cracks are formed, the more damage caused to the rock mass.

To minimize the power damage, there is a need to improve the soft roof rock mass stress state and reduce the development degree of cracks.

6. 5. 2　Mechanism for Failure of the Roof Rock Mass in Retaining Roadway

As shown in Figure 6.2(c), cross section C-C in the stabilized zone in retaining roadway experience a fracture deformation process at the working face end that can be influenced by the main roof rotary deformation and stress recovery process beside the working face, resulting in its larger damage and deformation degree, causing a weakening of its integrity. After strengthening the support of the roof rock mass in the retaining roadway and constructing the artificial filling wall body, their integrity and stability were improved and the bulking force was restored to a certain degree. For a certain retaining distance, after the strengthened support body was recycled, due to the artificial filling wall's higher strength and stiffness, the bulking force would release along its inside. Therefore, the soft roof of retained entry caved at the man-made constructed wall (shear failure line in Figure 6.10) due to forces such as the overburden pressure (P), the bulking stress (P_s), rock mass gravity (G) and the bolt, mesh, and cable support force (T), which formed a 'similar cantilever beam' structure, as shown in Figure 6. 10.

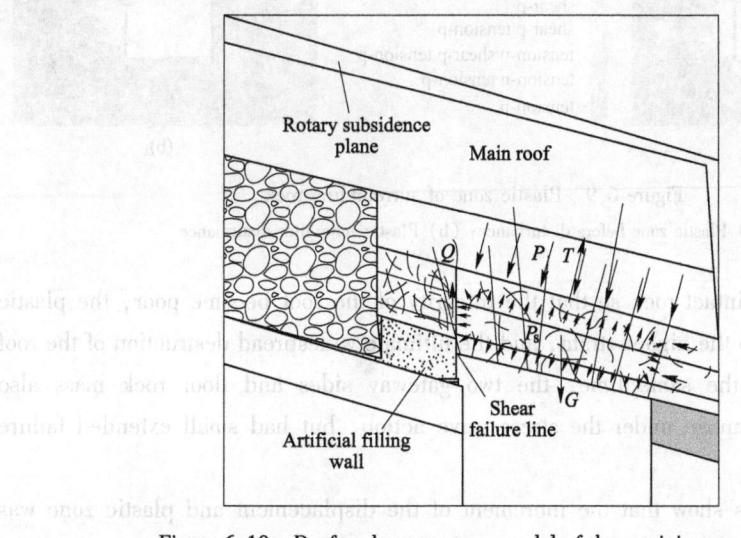

Figure 6. 10   Roof rock mass stress model of the retaining roadway

6. 6   Roof Support Countermeasures

To develop a supporting method for the roof rock mass at the working face end and retained gateway, the deformation and force in the roof rock mass must be analyzed. The rock bolt is the basic supporting tool. First, we calculated the 'cable' element to represent the rock bolts on the working face end with several different bolt installation angles in 3DEC model to find the bolt end displacement change characteristics. Then, we analyzed the bolt limit equilibrium tension force change features in the retained gateway roof rock mass with the change of the bolt installation angle though the force balance equation.

6. 6. 1   Deformation Analysis of Working Face End Roof

From the soft roof failure process and failure characteristics, the following mechanism for the soft roof support can be obtained:

(1) Providing the roof rock mass extrusion stress and changing the rock mass stress state to improve their strength[8].

(2) Preventing cracks from generating and propagating to increase both the strength of the affected roof strata and the stiffness of the whole bolted strata to reduce roof defor-

mationand dynamic damage.

(3) Reducing the broken rock mass bulking force and maintaining the stability of the roadway roof by supporting in time.

According to the above points and the 'bolt extruding reinforcement theory' by pretensioned bolts providing a compressive zone in the axial direction[8], the working face end soft roof mechanical structure should be similar to that in Figure 6. 11 with bolts supporting, so the installation angle is suggested to be $0° < \alpha < 90°$.

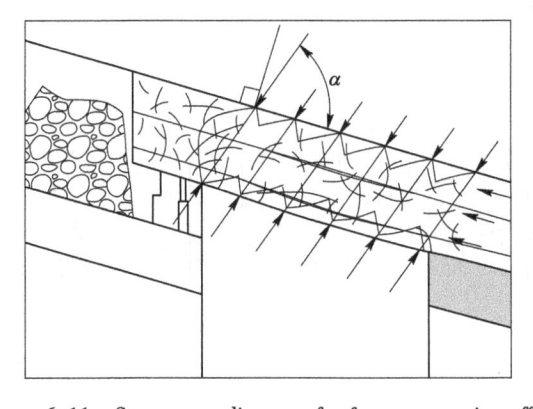

Figure 6. 11   Stress state diagram of soft top supporting effect

The distributed axial stress of the pre-tensioned bolts in the extension direction and the lateral stress vertical axial direction can alter the rock stress state, especially for the thin and weak immediate roof conditions[10]. Here, we use the same numerical model and the physical and mechanical parameters of the rock mass in Figure 6. 4 to analyze the effect of the bolt installation angle α on the stability of the roof rock mass. Bolt installation angles of 50°, 60°, 70°, and 80°, as shown in Figure 6. 12, were applied in the numerical model and the monitoring bolt end displacement was at $y = 45m$ with the same excavation process as shown in Figure 6. 4 conducted, and the results are shown in Figure 6. 13. As actual field bolt installation at an angle α in the mine roof close to 90°, similar to in Figure 6. 1(b), it was taken as 90° to facilitate the analysis.

Figure 6. 13 shows that:

(1) The monitored displacement had a good regularity and the displacement curve showed a trend of a 'concave' type with α increasing as a whole. The vertical displacement of roof was minimal when $\alpha = 60°$, and the displacement increased when $\alpha = 70°$ and 80°, but it had a smaller increment.

(2) For a certain α, the point displacement increased as the distance to the gob de-

· 98 · 6　Soft Roof Failure Mechanism and Supporting Method for Retaining Gateway

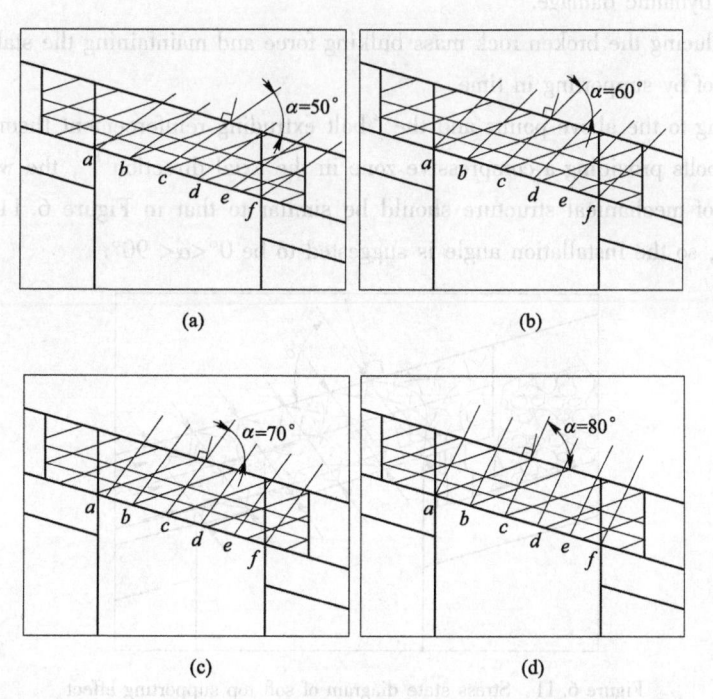

(a)　　　　　　　　　　　　　　(b)

(c)　　　　　　　　　　　　　　(d)

Figure 6. 12　Arrangement diagram of different bolt installation angles

(a) 50°; (b) 60°; (c) 70°; (d) 80°

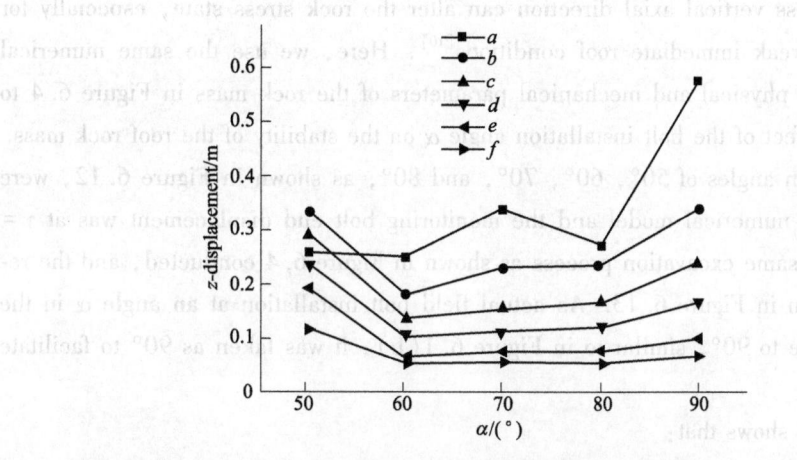

Figure 6. 13　Bolts end displacement

creased, that is to say, the roof rotated and sank.

(3) The displacement is larger under the current anchor installation angle ($\alpha = 90°$) condition. Its differences were 155mm and 321mm from the minimum displacement.

6.6.2 Bolt Limit Equilibrium Tension Force

It is known that the working principle of bolts is to maintain the roof rock mass stability in the early stage and to control roof deformation in the later stage. Moreover, carrying out gob-side entry retaining is a systematic project and the roof deformation in the later retaining stage, when maintaining the working face end roof rock mass stability should be considered. After the construction of the artificial filling wall and recycled reinforcement support body, the roof shear failure, as shown in Figure 6.2 (c), is mainly influenced by the bolt-cable-mesh force T, gravity G, shear forces Q_a and Q_b, and the lateral forces F_a and F_b [reference in Figure 6.6(c)], regardless of the overburden pressure because of the roof separation, as shown in Figure 6.14.

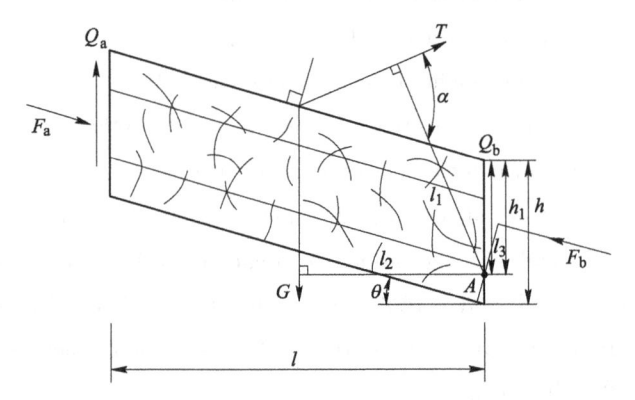

Figure 6.14 Stresses-Balanced schematic of roof rock mass

At the moment, the roadway roof rock mass undergoes serious failure and its integrity is poor, so it loses its rigid body properties. Therefore, there is distance h_1 between rotating point A and the upper endpoint, so it cannot have a certainty value under the action of torque and $0 \leqslant h_1 \leqslant h$.

On the other hand, the roof had been cut along the edge of the gob near the artificial filling wall, so the lateral pressure and shear force can be ignored. In the simplified calculation, it can be set that $F_a = 0$, $Q_a = 0$. The stress balance relationship in the layer strike direction and vertical layer direction are established, as well as the moment balance relationship established by point A, as shown as follows:

$$T\sin\alpha + Q_b\cos\theta - G\cos\theta = 0 \tag{6.1}$$

$$T\cos\alpha + G\sin\theta - Q_b\sin\theta - F_b = 0 \tag{6.2}$$

$$Tl_1 - Gl_2 - F_b l_3 = 0 \tag{6.3}$$

Where, l_1, l_2 and l_3 are the moment arms of T, G and F_b to point A, respectively, given by:

$$
\begin{cases}
l_1 = \sqrt{\left(h_1 + \dfrac{l}{2}\tan\theta\right)^2 + \dfrac{l^2}{4}\sin\left(\alpha + \arctan\dfrac{2h_1 + l\tan\theta}{l} - \theta\right)} \\[2mm]
l_2 = \dfrac{l}{2} \\[2mm]
l_3 = \left(h_1 - \dfrac{h}{2}\right)\cos\theta
\end{cases}
\tag{6.4}
$$

Where l is the roadway width 3. 8m, θ is the coal seam dip angle 16°, and the gravity G is calculated according to the following Equation:

$$
G = a(n-1)bh\bar{\rho}g\cos\theta
\tag{6.5}
$$

Where a, b——the bolt inter-row spacing, both 800mm;

n——the number of anchor bolts, $n=6$;

h——the study height of the roof rock mass, $h=1.5$m;

$\bar{\rho}$——the average density of the rock mass, $\bar{\rho}=3000$kg/m^3;

g——the gravitational acceleration, $g=10$m/s^2.

When $T=6\times90$kN, G is calculated by Equation (6.5). Assuming $Q_b=0$, it can be obtained that $\alpha=15°$ and 165° according to Equation (6.1). This means that, if α meets the condition of $15°\leqslant\alpha\leqslant165°$, the anchor tension force T is greater than or equal to the roof rock mass gravity in the vertical layer direction. Therefore, the roof can remain stable in this direction.

Furthermore, it can be observed from Equation (6.2) that α impacts the balance relationship of Equation (6.2) and determines the size of the lateral extrusion force F_b. If $0°<\alpha<90°$, it can not only satisfy the balance relation but also have a greater F_b value, as does the extrusion stress between the broken roof rock mass in the layer strike direction.

Substituting Equations (6.2), (6.4) and (6.5) into Equation (6.3) yields, we obtain:

$$
T = \frac{G(l_2 + l_3\sin\theta) - Q_b l_3\sin\theta}{l_1 - l_3\cos\alpha}
\tag{6.6}
$$

In reality, the lower roof stratum fails more seriously than the upper rock stratum, indicating that the roof rotating point A is usually above the roof surface. However, as it is impossible to determine the true position of the rotating A, we assume that $h_1=h$, implying that the rotating point A is at the bottom of study rock body. On the other hand, if the fracture roof rotated, the coal side shear force would decrease more, so we also assume $Q_b = 0$. Then, substituting each parameter of the mine mentioned above into Equation (6.6), produces:

$$\sin(\alpha + 31.1°) - 0.2584\cos\alpha = \frac{104.12\text{kN}}{T} \qquad (6.7)$$

By analyzing Equation (6.7), the following results are obtained:

(1) If the T value is small, the right value becomes greater than the left maximum value, so it cannot meet the balance relationship, and the bolts are not able to play their roles in supporting. This shows that when having a certain bolt installation angle α, there must be some tensile force T of the bolts to make the right value equal the left value in Equation (6.7). It also illustrates the importance of the bolts being pre-tensioned when used in the rock mass.

(2) When α takes the values of 50°, 60°, 70°, 80°, and 90°, the relation curve of the bolt tension T under roof rock mass ultimate stability conditions with an installation angle α is shown in Figure 6.15(a), and the change trend is similar to the curve in Figure 6.12.

Here, it can be explained that when α is kept constant, the smaller the T value, the smaller the force required for the roof rock mass limit equilibrium under the action of the torque; When the actual bolt tension values remain unchanged, the roof rock mass have a minimum displacement, which requires a minimal bolt tension force (T_{min}) for limit equilibrium. For example, in Figure 6.15(a), the largest displacement or the worst stability is when $\alpha = 50$ and the minimum displacement or the best stability is when $\alpha = 70°$ of the roof rock mass. Hence, the roof displacement is proportional to T.

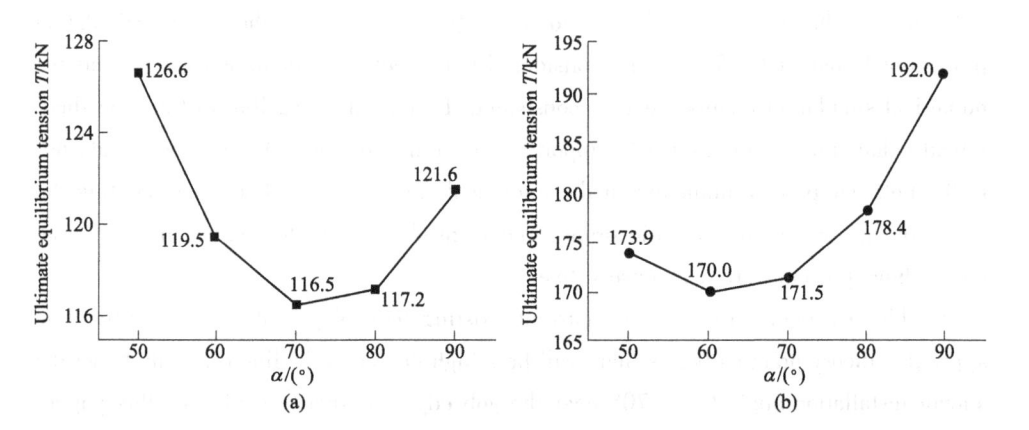

Figure 6.15　Relationship of limit equilibrium, anchor tension force, and installation angle

(a) The study condition when $\theta = 16°$; (b) Another condition when $\theta = 0°$

Combining Figures 6.13 and 6.15(a), it can be observed that when $\alpha = 70°$, it can provide a failure rock mass with extrusion pressure and alter the stress state, and it can also compress joints and fissures to reduce its opening and the dynamic damage at the same time. Hence, it is advantageous to control the working face end roof and maintain the roof stability in the gob-side entry retaining.

6.7 Discussion

The thickness of the soft roof stratum in a gently inclined coal seam is small and its strength is low, so mining activities can cause the roof rock mass to serious break and extremely easily cave in, producing a great threat to normal production activities and personnel safety. The selection of a pre-tensioned anchoring technique and bolt arrangement style can improve the bulking roof rock mass stress state by preventing cracks from expanding and reducing the roof separation. To meet the requirements of the retained head entry, a bolt arrangement form is proposed. Furthermore, the application of Equation (6.6) can be used for the discussion of the supporting measures under similar coal seam geological conditions of, for instance, different coal seam dip angles, gateway widths, and soft roof thicknesses. Because researchers provided more insights into the soft roof failure mechanism of horizontal coal seam by retaining gateway to set $\alpha = 90°$, thereby assuming the caving height $h = 2m$, coal seam dip angle $\theta = 0°$, other conditions remain unchanged as above, thereby determining the proper relationship concerning α and T, as shown in Figure 6.15(b). It shows that T is much large when $\alpha = 90°$ under the high thickness soft roof and horizontal coal seam condition.

On the one hand, the lateral shear forces of Q_a and Q_b in the theoretical calculation process of Equation (6.7) are not considered here, but the calculation results and the numerical simulation results are very consistent. Hence, it is feasible to treat the theoretical calculation results as a bolt support reference in the field. On the other hand, only the bolts support is mimicked in the numerical simulation, without incorporating the action of the individual hydraulic props, anchor mesh, and anchor cable, implying that the findings present here are conservative.

In addition, being restricted by current existing technology, it may be difficult to apply this theory practically, as there will be a high degree of drilling difficulty to set the anchor installation angle to $\alpha = 70°$ near the gob edge, as recommended by this paper. However, bolts with a theoretical value α can be installed on the coal side, and for the gob side bolts, α can be set as close as possible to the theoretical value so that they will not be influenced by the roof caving in near the gob, as shown in Figure 6.2, and to make the most use of the bolts resource by improving their force. The point of the roof failure angle decreased as the horizontal stress level increased, indicating that failure tends to occur around the entry corners when the horizontal stress was low[27]. However, the corner bolts (like e and f in Figure 6.12) that are installed tilted to the coal side can increase the horizontal stress, so they can prevent shear failure around the entry corners very well.

6.8 Chapter Summary

This chapter analyzed the roof rock mass failure characteristics and their failure mechanism in the working face end and the gob-side entry retaining, and discussed its support technology according to the special gently inclined seam occurrence conditions. The following conclusions were reached from the analysis process:

(1) After the working face advances, it is found that the horizontal stress of the soft rock mass at the working face end does not exhibit a large magnitude unloading on the coal wall side, but the horizontal stress momentarily fell, and a tensile stress appeared on the gob side. The vertical stress in the gateway central dropped significantly, almost down to zero.

(2) The sinking and separation of the soft roof rock mass in the gently inclined coal seam working face end is affected by the front abutment pressure and the hanging roof on gob side. The initiation and propagation of cracks and the fractures of the rock mass are produced by the actions of the lateral stress unloading loose, tensile, and shear stresses in the low layer caused by uneven support and no coupling support and dynamic disturbances.

(3) The roof rock mass failed in shear mode along the inside of man-made constructed wall in the stability zone of the retained gateway, due to the overburden pressure, bulking force, roof gravity, and combined supporting force. The failed roof forms a 'similar cantilever beam' structure.

(4) The equation of the bolt ultimate equilibrium tension force, a function of the seam inclination, gateway width, soft roof thickness, and bolt installation angle, was established according to the stress balance analysis of the roof rock.

(5) To prevent the working face end soft roof rock mass from increasing its deformation and becoming significantly fractured, and also to maintain the gob-side entry retaining roof's stability, it is suggested that the gateway roof bolt installation angle be 70° to provide an extrusion stress, change the rock mass stress state, and improve their strength for better entry maintenance.

References

[1] Zhang G H. Roof cracking reason analysis about gob-side entry retaining under initiative support [J]. J China Coal Soc, 2005, 30: 429-432.

[2] Gou P F, Zhang Z P, Wei S J. Physical simulation test of damage character of surrounding rock under different levels of the horizontal stress[J]. J China Coal Soc, 2009, 34: 1328-1332.

[3] Coggan J, Gao F Q, Stead D, et al. Numerical Modelling of the Effects of Weak Immediate Roof Lithology on Coal Mine Roadway Stability[J]. Int J Coal Geol, 2012, 90: 100-109.

[4] Zhang D S, Mao X B, Ma W D. Testing study on deformation Features of Surrounding Rocks of Gob-Side Entry Retaining in Fully-Mechanized Coal Face with Top-Coal Caving[J]. Chin J Rock Mech Eng, 2002, 21: 331-334.

[5] Xie W B. Influence factors on stability of surrounding rocks of gob-side entry retaining in top-coal caving mining face[J]. Chin J Rock Mech Eng, 2004, 23: 3059-3065.

[6] Nemcik J, Ma S Q, Aziz N, et al. Numerical modelling of failure propagation in fully grouted rock bolts subjected to tensile load[J]. Int J Rock Mech Min Sci, 2014, 71: 293-300.

[7] Ma S Q, Nemcik J, Aziz N. An analytical model of fully grouted rock bolts subjected to tensile load[J]. Constr Build Mater, 2013, 49: 519-526.

[8] Yang S S, Cao J P. Evolution mechanism of anchoring stress and its correlation with anchoring length[J]. J Min Saf Eng, 2010, 27: 2-7.

[9] Zheng X G, Zhang N, Xue F. Study on stress distribution law in anchoring section of prestressed bolt[J]. J Min Saf Eng, 2012, 29: 365-370.

[10] Cao S G, Zou D J, Bai Y J, et al. Surrounding rock control of mining roadway in the thin coal seam group with short distance and "three soft"[J]. J Min Saf Eng, 2011, 28: 524-529.

[11] Fan K G, Jiang J Q. Deformation failure and non-harmonious control mechanism of surrounding rocks of roadways with weak structures[J]. J China Univ Min Technol, 2007, 36: 54-59.

[12] Hua X Z, Ma J F, Xu T J. Study on controlling mechanism of surrounding rocks of gob-side entry with combination of roadside reinforced cable supporting and roadway bolt supporting and its application[J]. Chin J Rock Mech Eng, 2005, 24: 2107-2112.

[13] Yan Y B, Shi J J, Jiang Z J. Application of anchor cable with bolt and steel band coupling support technology in gob-side entry retaining[J]. J Min Saf Eng, 2010, 27: 273-276.

[14] Chen Y, Bai J B, Wang X Y, et al. Support technology research and application inside roadway of gob-side entry retaining[J]. J China Coal Soc, 2012, 37: 903-910.

[15] Jing L, Stephansson O. Fundamentals of discrete element methods for rock engineering-theory and applications[J]. Elsevier: Amsterdam, 2007.

[16] Gao F Q, Stead D. Discrete element modelling of cutter roof failure in coal mine roadways[J]. Int J Coal Geol, 2013: 158-171.

[17] Cundall P A, Hart R D. Development of generalized 2-D and 3-D distinct element programs for modelling jointed rock[J]. US army engineering waterways experiment station: Minneapolis, MN, USA, 1985, SL-85-1.

References

[18] Firpo G, Salvini R, Francioni M, et al. Use of digital terrestrial photogrammetry in rocky slope stability analysis by distinct elements numerical methods[J]. Int J Rock Mech Min Sci, 2011, 48:1045-1054.

[19] Gao F Q, Stead D, Kang H P, et al. Discrete element modelling of deformation and damage of a roadway driven along an unstable goaf——A case study[J]. Int J Coal Geol, 2014, 127: 100-110.

[20] Xie H P, Gao F, Ju Y, et al. Quantitative definition and investigation of deep mining[J]. J China Coal Soc, 2015, 40:1-10.

[21] Hou C J, Guo L S, Gou P F. Rock Bolting for Coal Roadway[M]. Xuzhou: China University of Mining and Technology Press, 1999.

[22] Kang H P. Study and application of complete rock bolting technology to coal roadway[J]. Chin J Rock Mech, 2005, 24:161-166.

[23] Zhang Y Q, Peng S S. Design considerations for tensioned bolts[C]. In proceedings of the 21st international conference on ground control in mining, West Virginia University, 2002.

[24] Lu W B, Yang J H, Yan P, et al. Dynamic Response of Rock Mass Induced by the Transient Release of in-situ Stress[J]. Int J Rock Mech Min Sci, 2012, 53:129-141.

[25] Felice J J, Beattie T A, Spathis A T. Face velocity measurements using a microwave radar technique[C]. In Proceedings of the 7th Research Symposium on Explosives and Blasting Technique, 1991, 2(6,7):71-77.

[26] Preece D S, Evans R, Richards A B. Coupled explosive gas flow and rock motion modeling with comparison to bench blast field data[C]. In Proceeding of the 4th International Symposium on Rock Fragmentation by Blasting, 1993,7(5-8):239-246.

[27] Peng S S. Coal Mine Ground Control [M]. 3rd ed. Beijing: China University of Mining and Technology Press, 2013.

7 The Failure Characteristics and the Supporting Technology for Pre-retaining Gateway

7.1 Introduction

At present, most of the design of basic support parameters of gateway is only aimed at the early stage of gateway driving and workface mining, without systematic consideration of the late stage of retaining gateway, which makes the support technology for retaining gateway lag behind or support cost greatly increased, and it is usually difficult to achieve the expected effect. Based on this understanding, on the basis of fully understanding the failure characteristics of surrounding rock of gateway under different geological conditions, the advanced support technology should be used to carry out the basic support in the driving stage and strengthen the support during the mining process for pre-retaining gateways.

The geological environment of gateways is complex, and most of them are sedimentary structures, and the bedding plane and rock mass fractures cause the discontinuity of the surrounding rock mass. In addition, due to the difference of engineering purposes, the limitations of many research methods on surrounding rock failure of gateways are highlighted.

To further develop our understanding of the failure characteristics of pre-retaining gateways surrounding rock, and to provide a referenced supporting method, the following works were conducted in the present study. First, investigation of test fields, which included examination of geological conditions, size and shape parameters of gateways, and physical and mechanical parameters of surrounding rock for 55 typical gateways. Second, test of the gateways' broken width by GPR, and analysis of a cross-section diagram of the excavation broken zone combined with the broken width and plastic zone width obtained by three-dimensional distinct element code (3DEC). Third, experimental studied on the improved material parameters of bolts in the laboratory and in the field to prove its ability to provide high pretension force. Fourth, evaluation of its presented support theory of an EBZ, and analysis of the initial load-bearing zone of gateway surrounding rock and the selection principle of bolt support design parameters. Finally, a referenced support method was provided for gateways.

7.2 Understanding of Surrounding Rock Failure and Its Support

Gateway failure range plays a guiding role in determining the support parameters of surrounding rock, and efforts along these lines have been discussed by many researchers[1-5]. Terms, such as different failure ranges, excavation damaged zone (EDZ), disturbed rock zone (DRZ), broken rock zone (BRZ), excavation disturbed zone (EDZ), and excavation influenced zone (EIZ), have been defined by many causes, and it has been noted that after excavating the gateway, the stress balance state of surrounding coal and rock mass would be broken, convert to a biaxial or uniaxial stress states from a triaxial stress state, and form concentrated stress around the gateway. Coal and rock mass have different failure ranges after influenced by redistributed stresses. It is indicated that the main distinction of different failure ranges is the difference in physical, mechanical, and hydraulic properties of the coal and rock mass[6,7]. Moreover, the properties related to strength and stiffness are especially different. Therefore, intact zone, excavation plastic zone (EPZ) and excavation broken zone (EBZ) are employed for the present study, as shown in Figure 7.1[4,5].

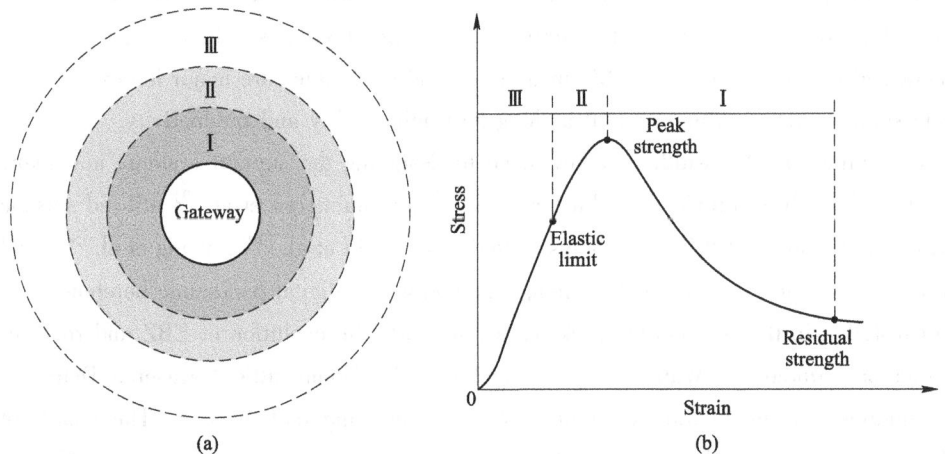

Figure 7.1 Surrounding rock failure mode of gateway

(a) Schematic diagram of failure zone;

(b) Failure zone in the view of stress-strain curve obtained by triaxial compression test

Ⅰ—Excavation Broken Zone(EBZ); Ⅱ—Excavation Plastic Zone(EPZ); Ⅲ—Intact Zone

The intact zone is the intact surrounding rock with an in situ stress state in which the properties of the rock masses have not been altered from the original state. The EPZ is defined as a zone without major changes in flow, transport properties that develop with micro-fracturing, and an integrated structure with a certain carrying capacity for deep coal and rock mass[6,7]. Thus, the EPZ is not the key zone to control in order to obtain stability of coal and rock mass that surrounds gateways. The EBZ is defined as a zone with major changes in flow and transport properties that develop with micro-fracturing; its size plays a significant role in the stability around man-made openings, and it has sparked deep discussion and research. Dong et al. [1] proposed the concept of EBZ and advanced the idea that the key factors to be considered for roadway support are the extent of EBZ and the bulking force caused by EBZ. Then, the assessment of EBZ was mainly used to design a support system that could maintain the stability of manmade openings. Therefore, the support design should be adjusted to utilize the self-bearing capacity of the surrounding rock masses. Wang et al. [4] investigated 14 typical roadway EBZs under both dynamic and static pressure and designed a supporting system. It was indicated that when the width of the EBZ was larger than the bolt length, the employment of bolts alone would be insufficient to maintain the roadway stability; in that case, reinforcement cables must be added to the support system. Tsang and Bernier[6] presented a number of studies that synthesized various ideas to identify the formation processes, parameters, and long-term safety state of EBZs in four types of geomaterials: crystalline rock, rock salt, and indurated and plastic claystones. Therefore, the assessment and characterization of EBZs in underground coal mines are major issues for roadway support system design as well as long-term mine safety and productivity.

Obtaining the EBZ width is a key stage to designing the support system, and many methods have been provided. Schuster et al. [8] and Malmgren et al. [9] utilized seismic waves to investigate EBZ in boreholes drilled in rock. Li et al. [10], Wang et al. [4], and Tan et al. [11] presented a borehole image method with a digital panoramic borehole camera installed in the surrounding rock masses to study the evolution of EBZ and roof deformation. Moreover, Wang et al. [4] combined the Nonmetallic Ultrasonic Detection techniques to evaluate the extent of EBZ in surrounding rock masses. The results of these studies showed that Borehole Camera Detection was an intuitive and an effective method.

On the other hand, to effectively evaluate the EBZ, some numerical models have been proposed. Hommand-Etienne et al. [12] and Golshani et al. [13] built the numerical model to analyze the plastic zone around man-made openings. They both concluded that the development of the EBZ was a function of time and that the shotcrete support was necessary to guarantee the stability of roadways and prevent further expansion and development of the damaged zone. Li[14] used a two-part liner elastic Hooke's numerical model in the TOUGH-FLAC3D code to study the mechanical response of the plastic zone. Pellet et al. [15] presented a 3D numerical simulation of the mechanical behavior of deep underground galleries with a special emphasis on the time-dependent development of the Excavation Damage Zone. Gao et al. [16] proposed a UDEC Trigon approach, and Kang et al. [17] adopted the method of intrinsic capability to simulate the initiation, propagation, and coalescence of fractures, as well as the interaction between them and any pre-existing discontinuities. In the UDEC Trigon approach, the EBZ can be observed clearly, and the supporting effect can be obtained obviously with a support system of bolts, cables and shotcrete combined with steel mesh and steel belts.

Although the above-mentioned approaches are imperative to evaluate the EBZ, they cannot be effectively used to predict all geological and geometrical conditions of man-made openings. For seismic waves, borehole images, and Nonmetallic Ultrasonic Detection techniques, boreholes would consume more time, money, and manpower. Particularly for the numerical model method, the results usually indicate that the width to the excavation boundary for the EPZ is larger than that of EBZ, as shown in Figure 7.1. Even for the UDEC Trigon approach, the results are significantly influenced by geological conditions and physio-mechanical parameters. Thus, the numerical results are sometimes to be referenced only for support.

To overcome these limitations, Ground Penetrating Radar (GPR) techniques are thusly proposed. GPR is a non-destructive method that provides a relatively quick geophysical measurement and is widely used for testing various engineering structures by providing continuous images of the interior of the media being analyzed[18,19]. GPR can provide information on the medium and details of a structure[20,21], such as pavement analysis[22], bridge and railway monitoring[23], the location of reinforcing bars and metal elements in concrete bases[24], the damage in reinforced concrete[18], and layer thickness[25]. GPR can also be adopted in coal mines. Church et al. [26] used GPR for strata control. Zhang et al. [27] utilized GPR to detect coal seam geological factors, such as fault structure, fracture zone, and collapse column. Strange et al. [28] described the application of GPR system for measuring coal thickness, coal depth, and near-surface interface in coal mining operations. Koarolu et al. [29] applied this method to obtain geological properties of coal seams near the surface.

During the process of using GPR, it was found that internal interference of hardware system influenced the result, such as jitter, bad antenna shielding effect, or more easily disturbed by coupling signal. Sometimes, it had weak reflection signals for some structures, and the technology for detecting weak signals needs improving. In addition, GPR was given priority for qualitative explanation at present, but quantitative standards remained to be further studied. However, GPR as an advanced non-destructive detection technology has the advantages of high precision, efficiency, and resolution, it is convenient to use and carry, and it provides fast results that are quite reliable. Therefore, GPR can be used to obtain crack distribution characteristics and broken widths in a test location, and its advantages are superior to other methods[30,31].

Moreover, the combined support technology mentioned above plays an important role in gateway EBZ control, and has realized that pretension force in bolts plays a decisive role. Because the pretension force bolt with the surrounding rock can form a 'stress arch' bearing structure, the active supporting role of bolt and the self-bearing capacity of the structure were given full play. Furthermore, Lang (1961) made a classic 'broken rock anchorage test', and vividly explained the bolt reinforcement effect. Therefore, research in the distribution rules of pre-stress in surrounding rock, the formation condition of compressive stress zones, the range of anchoring zones has an important theoretical significance and engineering application value, and many scholars contributed to this research direction. Gu et al. [32] studied the influence of pretension force and bolt length on the stress distribution zones around the bolt by a similarity simulation experiment. Kang et al. [33] summed up the engineering experience, a FLAC3D numerical simulation and similarity simulation method, and studied the support effect and action mechanism of pretension force bolt, the distribution form and influencing factors of pre-stress fields in surrounding rock. Lin et al. [34] analyzed the distribution characteristics of the pre-stress field of a single end anchorage bolt from a large model test, found that two concentration zones of compressive stresses formed near the two ends of free segments and a concentration zone of tensile stress formed near the anchorage segments. Guo et al. [35] researched the distribution of the stress field under the condition of end anchorage bolts with circular trays based on the elastic theory. Showkati et al. [36] studied the distribution of stress fields in the surrounding rock which contained vertical joints under the condition of end anchorage bolts. Ranjbarnia et al. [37] studied the distribution of stress fields in a circular roadway with pretension force bolt by full length anchorage, and analyzed the influence of pretension force and bolt density on that stress field. Ding et al. [38] and Wang et al. [39], useing a FLAC3D numerical simulation, studied the stress distribution characteristics in surrounding rock supported by pretension force cable. Wei and Li[40] adopting a FLAC3D numerical simulation studied the formation factors of the anchorage body in the use of pretension force bolt and the destabilization mechanisms resulting from the in-situ stress.

Many scholars, using diverse methods, researched stress fields, their influencing factors, and obtained numerous achievements. However, many experiments were conducted by using several bolts or a single bolt with low pretension force. Kang et al. [41] pointed out that the bolt pretension force range is usually between 30kN to 90kN at present, and when pretension force exceed 100kN, the surrounding rock could obtain a good support effect. With further understanding of the anchor action mechanism and the continuous improvement of bolt material parameters and support technologies, many new insights could be discovered.

7.3   Investigation of Test Field

The investigation fields are distributed over a large area, and the test coal mines are located in five mining areas in Sichuan Province, China; In total there are 55 typical gateways at 19 coal mines. The investigated factors includes the geological conditions, physical and mechanical parameters of surrounding rock, and gateway cross-section shape and size. Some of these features have a significant influence on the broken width, and some are for the numerical analysis.

7.3.1   Geological Conditions of Gateways

The geological conditions around the extracted gateway mainly include the lithology and thickness of surrounding rock, buried depth, dip angle, and thickness of the coal seam. Results indicated that it has complex geological conditions affected by severe geological movements. The key factors for influencing the broken width had large variable ranges, as shown in Table 7.1.

Table 7.1   Geological conditions parameters of test gateways.

Conditions	Buried Depth/m	Dip Angle/(°)	Thickness of Immediate Roof/m	Thickness of Coal Seam/m
minimum	approximately 200	9	less than 1	0.5
maximum	approximately 700	67	approximately 18	close to 5
most	300~500	25~40	3~6	0.8~2.5

On the other hand, we found that the coal-bearing strata of the tested gateways are Triassic and Permian, including the Xujiahe Formation (T_{3xj}), the Daqiaodi Formation (T_{3d}), the Longtan Formation (P_{2l}), and Xuanwei Formation (P_{2x}), as shown in Table 7.2. Most of the strata were formed under sedimentary and tectonic movement with a low cementation degree and poor mechanical properties of each rock material and bedding plane. In addition, the majority of the immediate roofs of the gateways were soft mudstone and argillaceous siltstone, with some of the siltstone——having poor vibration resistance——easily loosened or broken.

Table 7.2   Coal-bearing strata of tested gateways

Phanerozoic	Mesozoic	Triassic (T)	T_3	Xujiahe Formation (T_{3xj})	—
				—	Daqiaodi Formation (T_{3d})
	Paleozoic	Permian (P)	P_2	Longtan Formation (P_{2l})	Xuanwei Formation (P_{2x})
				—	

To conduct the numerical analysis by 3-Dimensional Distinct Element Code (3DEC)[42], the thickness of the roof, floor, and coal seam, as well as physical and mechanical parameters (e. g. , bulk modulus, shear modulus, density, internal friction angle, cohesion, and tensile strength) of related coal and rock mass were collected from each mine. The main material parameters of lithology are listed in Table 7.3. In addition, the mechanical and physical properties of the contacts between every two layers are described in[43].

Table 7.3   Material parameters of main lithology

Main Lithology	Density /kg · m^{-3}	Bulk Modulus K/GPa	Shear Modulus G/GPa	Cohesion C/MPa	Friction Angle $\varphi/(°)$	Tensile Strength σ_t/MPa
Coal	1400	2. 05	1. 02	1. 70	35	1. 1
Mudstone	2541	13. 3	9. 81	2. 8	42	2. 9
Sandstone	3020	13. 9	10. 4	6. 3	40	3
Siltstone	2600	2. 91	1. 04	1. 1	12	0. 3
Clay rock	2460	3. 98	2. 17	1. 8	25	1. 0

7.3.2   Cross-Section and Size

The investigation showed that most tested gateways used blasting excavation, few used comprehensive mechanized excavation, and a total of six kinds of cross-section shapes were collected from the 55 gateways, as shown in Figure 7.2, including the width and two side wall heights of each gateway for numerical analysis.

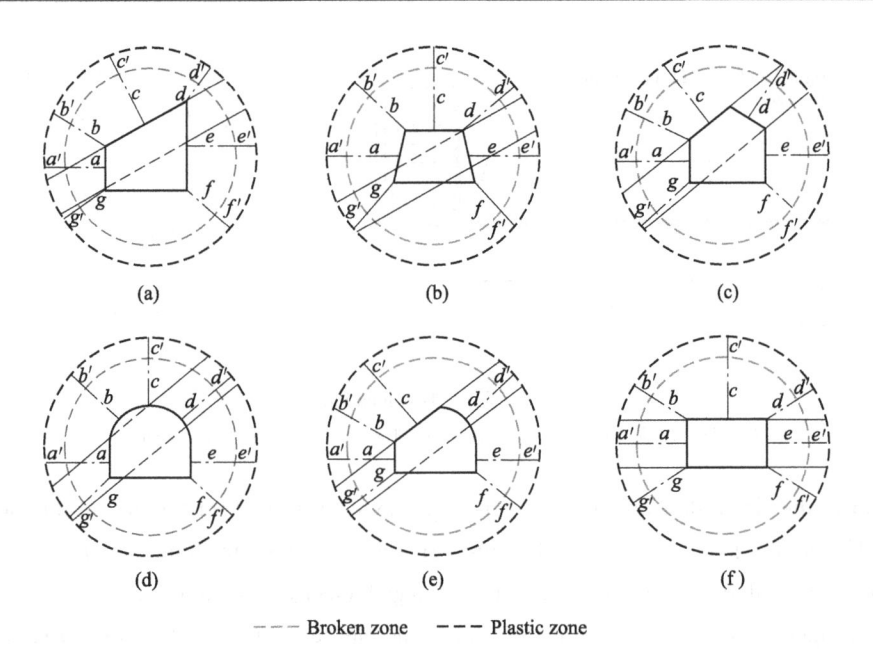

—— Broken zone   — — — Plastic zone

Figure 7.2   Gateways cross-section shape in investigate field
(a) Trapezoid-1; (b) Trapezoidal-2; (c) Special; (d) Arch; (e) Inclined arch; (f) Rectangle

7.4   Test of Gateways Broken Width

To obtain the result of broken width around gateways, the selection of test location, test principle, and equipment are imperative.

7.4.1   Test Location of Gateways

The test locations were in advanced roadways of head entry or tail entry beyond the front abutment pressure zone and in the original rock stress zones and under static pressure, where the roof and two side walls did not have large deformation and did not need to have reinforced support. The distances to the mining faces were usually greater than 30m, as shown in Figure 7.3(a). Each gateway cross-section was arranged with three survey lines on the high side wall, low side wall, and roof, and the maximum broken width in a survey line was selected as the utilization data (such as 'b' on the roof) when the conditions of the roadways were variable, as shown in Figure 7.3(b).

7.4.2   Test Principle of GPR

GPR technology has a similar principle to seismic wave and sonar technology, which launch high-frequency short-pulse electromagnetic waves into the rock medium; the

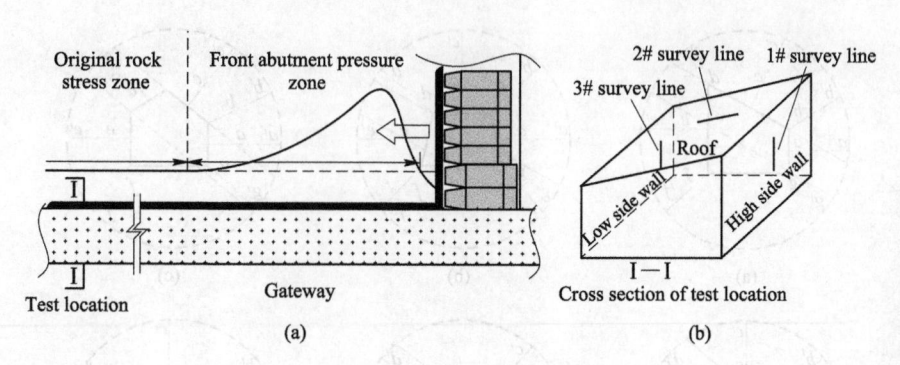

Figure 7.3   Test location

(a) The test location in the gateway; (b) The cross-section of test location.

spread of the signal depends on the high-frequency electrical characteristics of the medium. Generally, joint, crack, and fracture structures in a rock medium can change some of the electrical characteristics, and the changed electrical characteristics will cause electromagnetic wave signal reflection and generate radar reflected waves. Reflected waves will be received, magnified and digitized by the probe and stored in a computer. After editing the acquired data, we can obtain different types (e. g., waveform, grayscale, colour) of geology radar profiles. Thus, the measured results are obtained[31]. The working process and principle are shown in Figure 7.4(a). The depth of reflection interface can be obtained from:

$$z = \frac{\sqrt{v^2 t^2 - x^2}}{2} \tag{7.1}$$

Where z——the depth of the reflection interface;

t——the propagation time from the probe to the reflection interface of the electromagnetic wave;

x——the distance between the transmitter and the receiver;

v——the electromagnetic wave velocity in the medium.

Rock and coal mass is considered to be a low-loss medium. therefore, it is useful to approximate the wave velocity as[43]:

$$v = \frac{c}{\sqrt{\varepsilon}} \tag{7.2}$$

Where c——the wave velocity in a vacuum of approximately 30cm/ns;

ε——the relative permittivity of the medium.

The relative permittivity constants of some mediums are shown in Table 7.4. The medium of separation is usually air. It can be observed from the table that the relative per-

7.4 Test of Gateways Broken Width   · 115 ·

mittivity of the rock and coal medium, and the separation structure such as a fracture or crack, are rather different. GPR electromagnetic waves reflect well from the structure surface. Thus, detection of a failure structure in the roof and two side walls of a gateway by GPR is practicable[44].

Table 7.4 Relative dielectric constants of some mediums

Mediums	Air	Water	Limestone	Coal	Sandstone	Shale	Mudstone	Sandy Mudstone	Clay Rock
ε	1	81	7	4.5	4	5~15	5~25	5.53	8~12

In reality, the distance between the transmitter and the receiver is very short, so x is approximately equal to 0. According to Equations (7.1) and (7.2), Equation (7.3) was computed to determine the depth of the reflection interface:

$$z = \frac{vt}{2} = \frac{ct}{\sqrt{\varepsilon}} \qquad (7.3)$$

In this paper, an SIR-20 Model multichannel perspective radar made by GSSI in the USA, as shown in Figure 7.4(b), was adopted to investigate the broken widths of gateways. This radar is a type of highly efficient geophysical exploration instrument, especially for shallow detecting. A 100MHz centre frequency antenna was employed during the radar data acquisition.

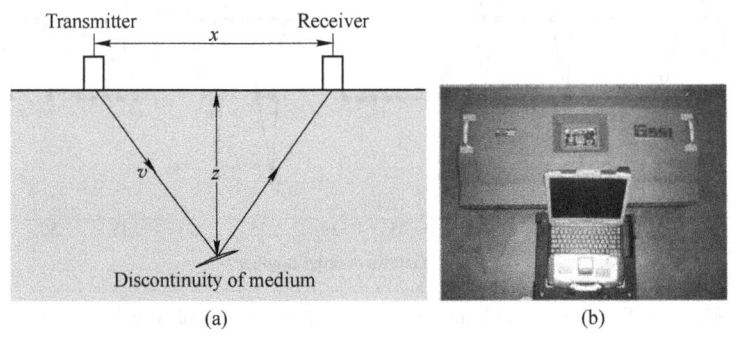

(a)   (b)

Figure 7.4 Test of ground-penetrating radar (GPR)

(a) The measuring principle of electromagnetic wave; (b) SIR-20 Model multi-channel perspective radar.

7.4.3 Results and Analyses of Gateways Broken Width

After four months of working with GPR, it yielded 55 typical gateways broken widths, including those in the high side wall, low side wall, and roof. In order to analyze the relationship between the broken width and the gateway geological conditions, the gateways were numbered from 1~55, as shown in Figure 7.5.

· 116 · 7 The Failure Characteristics and the Supporting Technology for Pre-Retaining Gateway

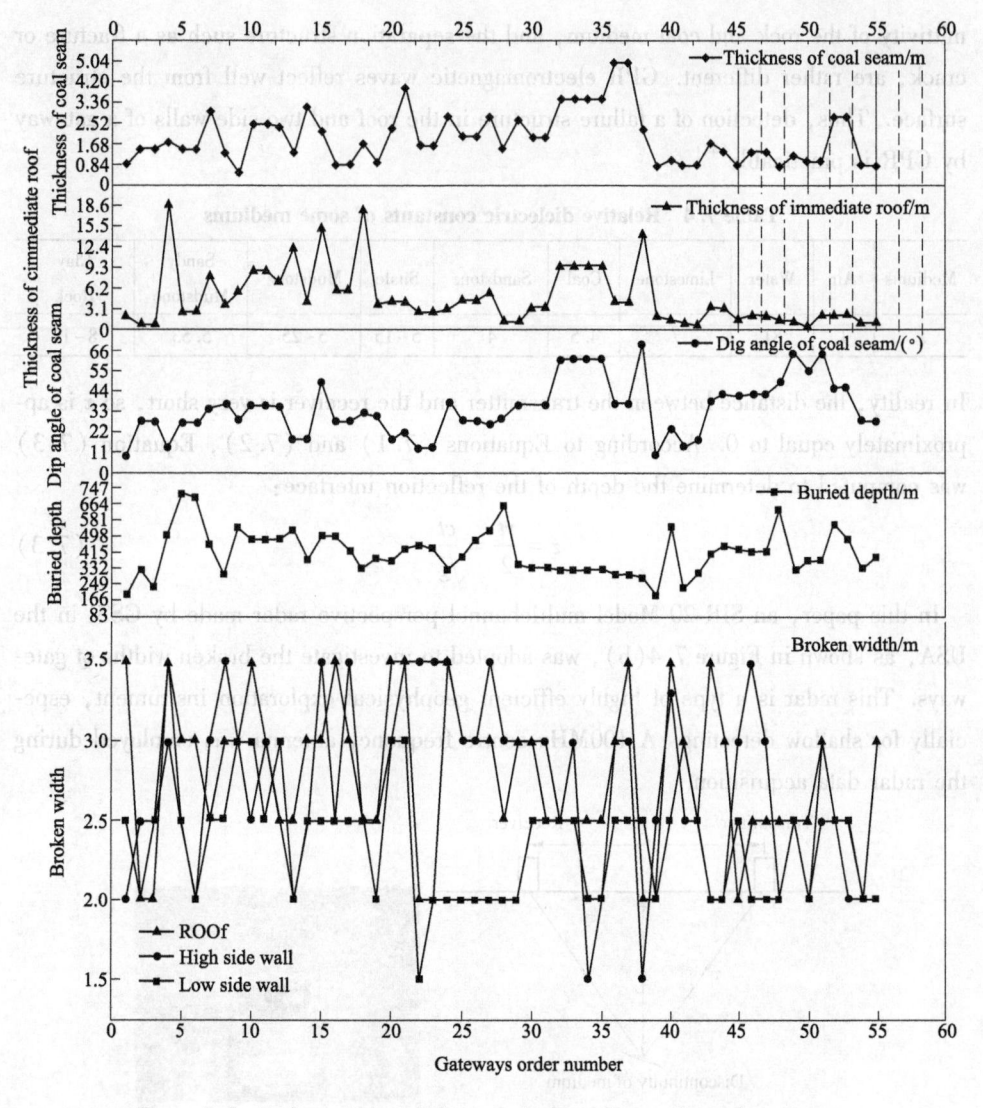

Figure 7.5 Broken width and the geological conditions of tested gateways.

The following can be observed from the broken widths:

(1) The broken widths of the tested gateways were large, the minimum value was
1.5m, and the maximum value was 3.5m.

(2) The broken width in the roof and high side wall were generally greater than that
in the low side wall.

Also, combined with the broken width, buried depth of the gateways, dip angle, and
thickness of immediate roof and coal seam, the following conclusions were obtained:

(1) There was a certain relationship between the thickness of the coal seam and the

thickness of the immediate roof for tested gateways with similar a change tendency, particularly between gateways No. 20~50.

(2) For the relatively larger thicknesses of the coal seam and immediate roof, the broken width in the roof and the high side wall were larger. Conversely, they were relatively small, which was more obvious among the gateways No. 35~55. The reason this phenomenon occurred was that the coal seam and roof rock mass had low strength and could easily become shear, causing tensile failure and increased failure range, and thus resulted in larger broken width.

(3) For coal seams with a small dip angle, the broken widths in the roof were larger, as seen in gateways No. 15~25. For a large dip angle of the coal seam, the broken widths in the roof were smaller, which was evident in gateways No. 30~40 and No. 45~55. The primary reasons lay in the large dip angle, the roof having transitioned from coal mass to rock mass, as well as the increased compressive stress and reduced tensile stress in the roof rock mass which led to small failures. At the same time, the small failure roof would be carrying a large load, leading to concentrated stress that could not be transferred into the deep rock mass to damage them. Thus, the above phenomenon would result.

(4) The smaller the buried depth, the smaller the broken width, which was obviously reflected among gateways No. 30~40. For gateway No. 40, the buried depth was relatively large, as was the broken width value. For a few gateways in front of and behind gateway No. 40, they had small buried depths and small broken widths. Explanation of this phenomenon is increased stress with increased buried depth, and increased concentration stress in the stress redistribution process. In addition, the tensile stress in the surrounding coal and rock mass would increase, and rock mass would fail more easily, eventually causing serious failure in the rock surrounding the gateway and increasing the broken width. Conversely, it decreased.

It can be observed from the above analyses that there were certain relationships between broken widths and the gateway geological conditions; but for some gateways, they were not quite clear. Through in-depth analysis, we found that the factors affecting the broken width changed simultaneously; they could not meet the single factor condition and thus were analyzed by single factor variation. For example, as the buried depth increased, the thickness of the coal seam decreased; meanwhile, the influence of the two factors on the change tendency of the broken width was not obvious, as remarkable seen in gateways No. 5~10. On the other hand, the blasting excavation and the geologic

· 118 · 7 The Failure Characteristics and the Supporting Technology for Pre-Retaining Gateway

structure could also lead to this phenomenon.

In reality, gateway broken width was controlled by the combined action of various factors, including initial stress (buried depth and tectonic stress), rock mass strength, gateway cross-section and support, and so on. The data of broken width in Figure 7.5 fully illustrated the combined action. Furthermore, broken width as a multi-factor single index to evaluate the damage degree of surrounding rock, and Figure 7.5 sufficiently demonstrated the practicability and scientificity of the evaluate index.

7.5 Cross-Section Diagram of Excavation Broken Zone (EBZ)

The excavation broken zone around gateways can present failure characteristics of the coal and rock mass and provide a strong pertinent support system for the roof and two walls according to the failure characteristics, rather than utilizing a conventional support system. However, the above field-tested broken widths showed that there were only three sets of data in a gateway cross-section, which cannot obtain the accurate EBZ. In addition, significant differences in cross-section shape, geological conditions and physical and mechanical properties of coal and rock mass between tested gateways would increase the errors of the EBZ. As a result, this paper acquired 7 directions of typical plastic width data by theoretical calculation in 3DEC and combined them with field-tested data to speculate the tested gateway broken width in 7 directions of the surrounding rock to obtain a high-precision EBZ.

7.5.1 Steps for Obtaining EBZ

The specific steps for obtaining EBZ were as follows:

(1) Build 55 plane strain models of 3DEC. Models with dimensions of 50m×1m×50m in the x, y, and z directions, respectively, were built by 3DEC according to each gateway geological condition, and one gateway model was shown in Figure 7.6(a). The physical and mechanical parameters of each mode material were obtained from the investigation (shown in Section 7.3.1) and obeyed the Mohr-Coulomb shear failure criterion and tensile failure criteria, and the following yield criterion, as shown in Equations (4) and (5), was used. The contact obeyed the Coulomb Slip model[42].

(2) Determine the boundary conditions of models. The state of in situ stress was defined by $\sigma_z = \lambda H$ in the vertical direction and $\sigma_x = \sigma_y = k\lambda H$ in the horizontal direction, where $\lambda = 0.027$ MPa/m, H was the buried depth, and k was the coefficient of horizontal stress obtained from each mine. A vertical pressure of $p = \lambda H$ was applied on the top

7.5 Cross-Section Diagram of Excavation Broken Zone (EBZ)   · 119 ·

surface, the velocity of the bottom surface was restricted in all three directions ($v_x = v_y = v_z = 0m/s$), and of the other four surfaces were restricted in the normal direction ($v_n = 0m/s$) as shown in Figure 7.6(a). First, it had an initial equilibrium calculation with elastic stress state, and after that gateway cross-section was excavated. Then, it had another equilibrium calculation and the plastic zone appeared around gateway.

(3) Obtaining plastic zone widths in 7 directions. The gateway broken widths in 7 directions were a, b, c, d, e, f, and g, while a, c and e were the field test values among them. After calculating, the 7 directions plastic zone width of a', b', c', d', e', f', and g' were extracted, as shown in Figure 7.2. One gateway of plastic zone widths is shown in Figure 7.6(b).

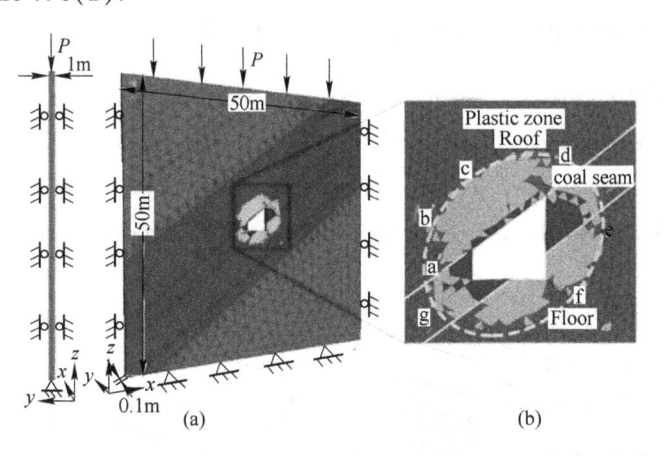

Figure 7.6   One gateway three-dimensional distinct element code (3DEC) mode
(a) Numerical model; (b) Plastic zone

(4) Calculating EBZ widths. The fractures and cracks of coal and rock in the excavation broken zone were attributed to the discontinuous medium and the lack of a mature theory to calculate the broken widths. On the other hand, the plastic zone in the numerical model was computed based on a continuum theory with failure criteria, which did not consider the joints behavior inside coal and rock mass. Therefore, EBZ was fundamentally different from the plastic zone; in general, the broken width was less than the width of the plastic zone ($a<a'$, $b<b'$, $c<c'$, $d<d'$, $e<e'$, $f<f'$, $g<g'$), and could not simply replace the EBZ with the plastic zone. However, to date, no quantitative relationship between the EBZ and the plastic zone has been found; this problem still requires significant work. Therefore, this paper treated plastic zone width, such as $a' \sim g'$, as the intermediate variable and calculated the EBZ width in another 4 directions according to a certain proportion as shown as follows:

$$
\begin{cases}
b = \dfrac{\dfrac{b'}{a'}a + \dfrac{b'}{c'}c}{2} \\[4mm]
d = \dfrac{\dfrac{d'}{c'} \times c + \dfrac{d'}{e'}e}{2} \\[4mm]
f = \dfrac{f'}{e'}e \\[4mm]
g = \dfrac{g'}{a'}a
\end{cases}
\tag{7.4}
$$

7.5.2　Cross-Section Diagram Resulting from EBZ

Each gateway EBZ cross-section diagram was mapped using a smooth curve after obtaining the other 4 directions of broken width values utilizing Equation (7.4). Because 55 cross-section diagrams were a large quantity, only 21 representative diagrams were listed, as indicated by the red elliptical shapes shown in Figure 7.7.

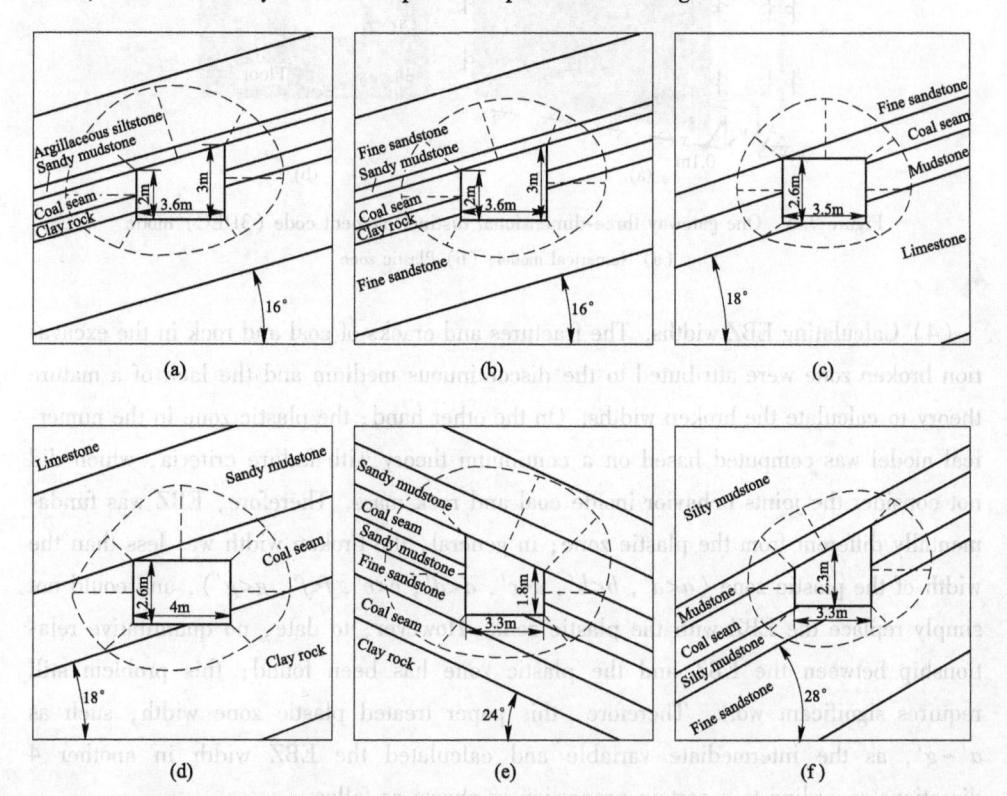

(a)　　　　　　　　　　(b)　　　　　　　　　　(c)

(d)　　　　　　　　　　(e)　　　　　　　　　　(f)

7.5　Cross-Section Diagram of Excavation Broken Zone（EBZ）　·121·

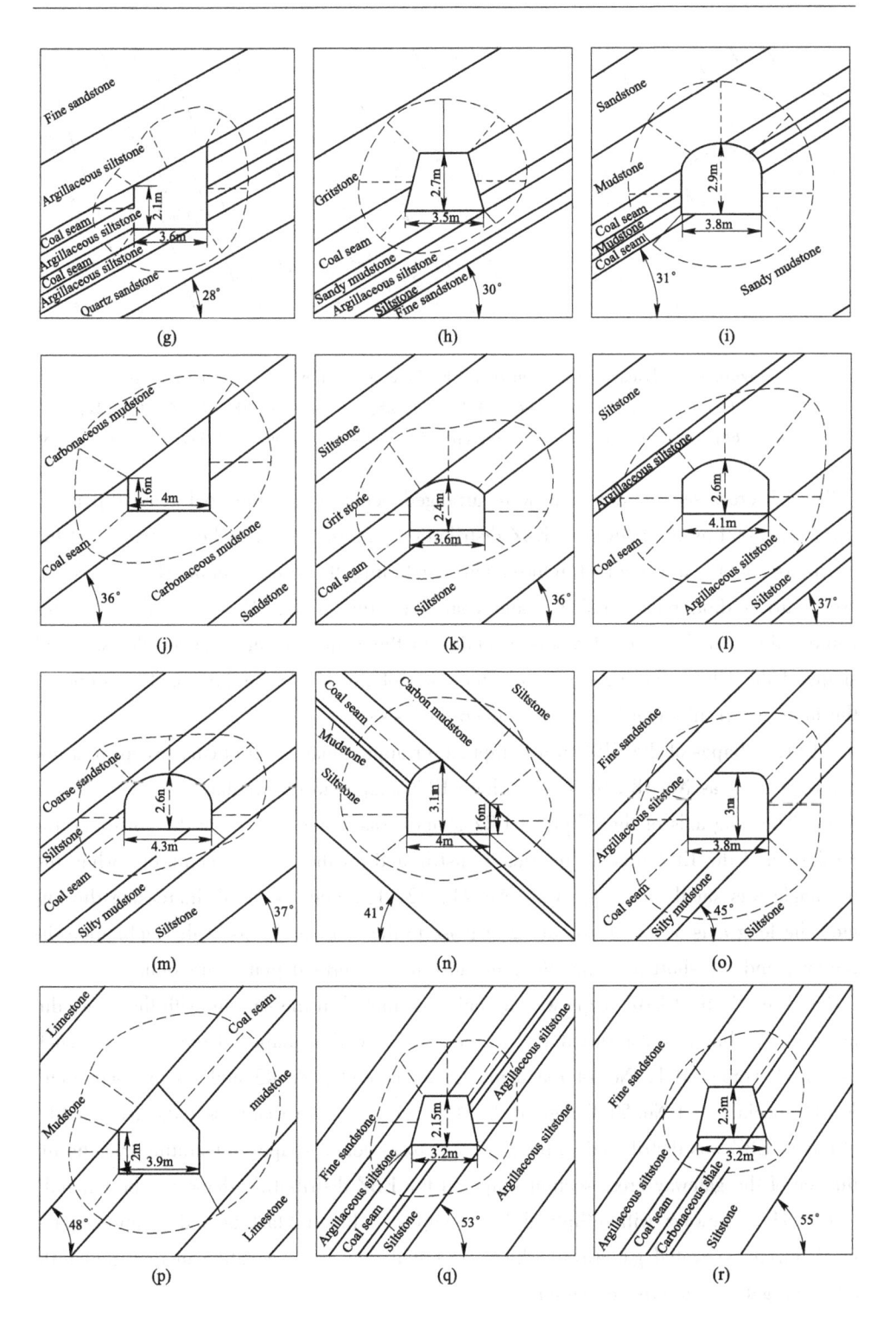

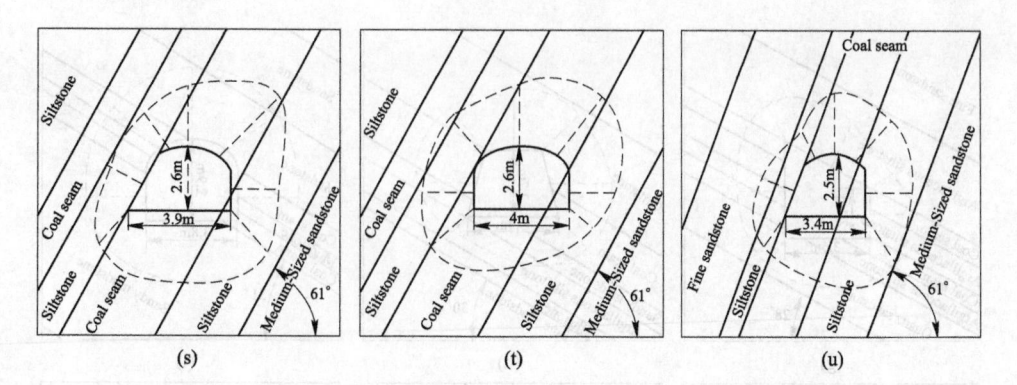

Figure 7.7　Excavation broken zone (EBZ) cross-section diagram for gateways

(a) 41#; (b) 42#; (c) 13#; (d) 14#; (e) 40#; (f) 2#; (g) 54#; (h) 18#; (i) 5#; (j) 12#;
(k) 31#; (l) 36#; (m) 29#; (n) 43#; (o) 53#; (p) 15#; (q) 48#; (r) 50#; (s) 34#; (t) 35#; (u) 38#

The 21 cross-section diagrams were arranged according to the coal seam dip angle size, representing 55 gateways' EBZ distribution patterns. It could be observed that the diagrams contain a wealth of information, including the cross-section shape, section size, surrounding rock lithology, and coal seam dip angle. Moreover, the diagrams could reflect whether the EBZ was extended to the main roof and convey the size and shape of the EBZ resulting from the interaction of the above conditions. Furthermore, the following results can be found by deep analysis:

(1) The shapes of the EBZs were elliptical or approximately elliptical, except in a few gateways such as No. 18 and No. 5, whose EBZ shapes were circular.

(2) The long axis of the elliptical EBZ shapes was along or close to the horizontal direction, and the EBZ was symmetrically distributed on the gateway centreline when the dip angle was small, as in gateways No. 41, 42, 13, and 14. With increasing dip angle, the long axis of the ellipse along or close to the coal seam was in the inclination direction, and the short axis was along or close to the vertical plane direction.

(3) The elliptical EBZ shape was not only distributed in the stratum with the entire dip angle range, but was also distributed in the 6 cross-section shapes of the gateways, such as No. 41 Trapezoid-1, No. 14 rectangular, No. 31 arch, No. 43 and 34 inclined arch, No. 15 special, and No. 50 Trapezoid-2. These cross-section diagrams indicated that the elliptical EBZ was distributed without a fixed cross-section shape, illustrating that the influence of the gateway cross-section shape on the EBZ distribution shape can be ignored.

(4) The gateways with elliptical EBZ shapes included blasting and comprehensive mechanized excavated gateways, which presented that the excavating method had little effect on gateway failure formation.

7.5.3 Failure Characteristics of Gateways Surrounding Rock

The EBZ shapes for the 55 gateways above provided a preliminary understanding, but we can infer that other gateways in mines should have these characteristics. Followed is the detailed summary.

For a small coal seam dip angle (α), a symmetric ellipse EBZ was distributed at the centre point A of the gateway $(a_1 = a_2)$ in the long axis direction, as shown in Figure 7.8(a). As α increases, asymmetric ellipse EBZ was distributed to point A $(a_1 > a_2)$, resulting in serious failure of coal and rock mass in the high side wall, as seen in Figure 7.8(b). In addition, in the short axis direction of the EBZ, $b_1 > b_2$ was indicated in above two cases.

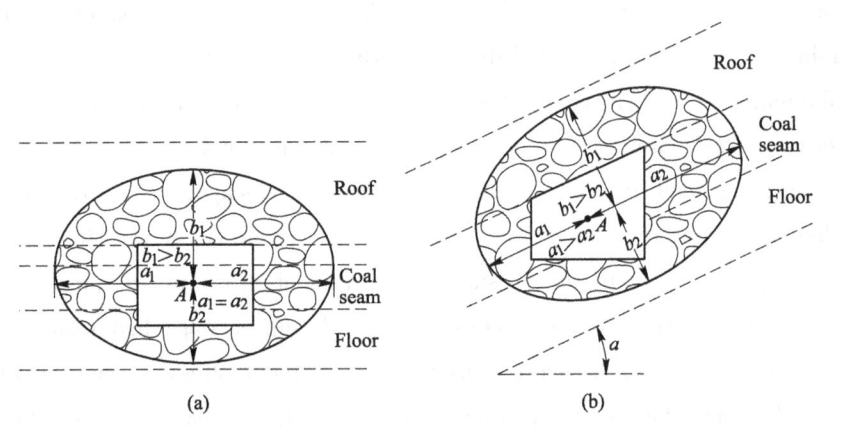

Figure 7.8　EBZ distribution characteristics
(a) For small coal seam dip angle; (b) For large coal seam dip angle.

7.6　Broken Width in Gateways with Different Excavation Method

Here, it can be seen that the model indicated in Figure 7.1(b) was associated with strain softening behaviour, and the numerical model used Mohr-Coulomb elastic parameters to calculate the plastic zone, which seems inconsistent. In reality, Figure 7.1(b) presents a typical failure zone, which has obvious plastic. However, brittle fracture for several rocks indicated that it had a small plastic zone, and could even be ignored. Thus, adopting the entire model according to the strain softening model was unreasonable. Also, the failure criterion of Mohr-Coulomb is a plane mode, which does not consider the second principal stress, and made it unsuitable for calculation with three-dimensional stress. But since no failure criterion with three-dimensional stress is widely used in numerical simulation process at present, Mohr-Coulomb model can be adopted in the underground engineering field, and the physical and mechanical parameters for the material can be easily collected because they are widely used. As is well-known, the plastic zone formed was just when the stress went beyond the yield point, and the broken zone formed when the stress went beyond the peak point in the stress-stain curve, so the plastic width was larger than broken width, as analysis in Section 7.5.1 covered, and we employed the proportional relationship of plastic zone width with broken width. Moreover, many gateways were excavated by blasting and its damage couldn't be modeled by the strain softening model. Thus, this method of utilizing the plastic zone width of Mohr-Coulomb model to calculate the broken width in the other 4 directions was feasible.

In addition, excavation methods had a great influence on the distributions of stresses and influenced the resulting EBZ distribution. Blasting excavation included two stages: blast forming stage and stress adjustment stage[45-50]. According to the blasting mechanism of rock, blast forming stage occurs over a short time, and can be divided into dynamic effect stage and static effect stage. In dynamic effect stage, stress wave and strong unloading effect after detonation effect would produce tensile stress, and lead to tensile fracture in rock. In static effect stage, the residual gas pressure produced after detonation effect works into the cracks. Stress adjustment stage had relatively long time——the failure being basically caused by concentration stress——and the gateway excavated by comprehensive mechanized methods usually only has the stress adjustment stage. Therefore, blasting excavation had larger broken width than that for comprehensive mechanized excavation.

Table 7.5 lists the broken width, the geological conditions, and excavated method for 4 gateways. Two gateways were located in the same coal mine area, and the other two were in the same coal mine. It could be seen that 5102 South tailgate and 3102 headgate had similar geological conditions, gateway cross-section shape, and size. However, 3102 headgate, excavated by blasting, had a larger broken width in low side wall, roof, and high side wall than that of 5102 south tailgate excavated by comprehensive mechanized methods. Similarly, 2121-31 tailgate and 2115 No.3 headgate had the same geological conditions and gateway cross-section parameters, but 2115 No.3 headgate, excavated by blasting, had a larger broken width in low side wall, roof, and high side wall than that of 2121-31 tailgateexcavated by comprehensive mechanized. It was fully verified that blasting had a more damaging effects to rock.

7.7 Experimental Studies on the Improved Material Parameters of Bolts · 125 ·

Table 7.5 Broken width in gateways with different excavation method

Coal Mines		Gateway	Section-Cross	Excavation Method	Bottom Width /m	Dip Angle /(°)	Broken width/(°)		
							Low Side Wall	Roof	High Side Wall
The same coal mine area	Lizi Ya	5102 south tailgate	Special	comprehensive mechanized	3.5	40	2.42	2.5	2
	Lizi Ya south two	3102 headgate	Special	blasting	3.9	48	2.5	3	3.5
Dabao Ding		2121-31 tailgate	trapezoid	comprehensive mechanized	3.8	30	2.5	2.5	2.5
		2115#3 headgate	trapezoid	blasting	3.8	30	3	3.5	3

7.7 Experimental Studies on the Improved Material Parameters of Bolts

For the EBZ support, many scholars[49,50] believed that bolt anchorage length and the amount of bolt pretension force play an important role in controlling the surrounding rock. Kang et al. [41] pointed out that when pretension force exceeds 100kN, the surrounding rock can obtain a good support effect. To conduct an experimental investigation to test whether the mechanical properties of bolts, such as tensile strength and anchorage force, can provide high pretension force which is more than 100kN, it must first disscuss the anchorage length of a bolt and its mechanical effects.

7.7.1 Mechanical Effects of a Bolt with Pretension Force

After theoretical research, the results of a boltpretension force mechanical effects were obtained in Figure 7.9[49,52-56]. It suggested that:

(1) The bolt in the anchored segment has a peak shear stress and rapid decay after peak value and the increase of the pretension force cannot lead to the pretension force in an anchored segment spreading to deeper surrounding rock.

(2) The increase of the pretension force can improve the pretension force peak in an anchorage body and compressive stress on the surface of the surrounding rock.

(3) The bolt's axial force (pretension force) in the free segment remains about the same, but there is a sharp decay in the anchored segment, unable to spread to a deeper bolt. It means the blot pretension force almost works on the free segment of rock mass

and had little effect on an anchored segment (anchorage length of bolt) of rock mass. Thus, the anchorage length of a bolt should be carefully determined.

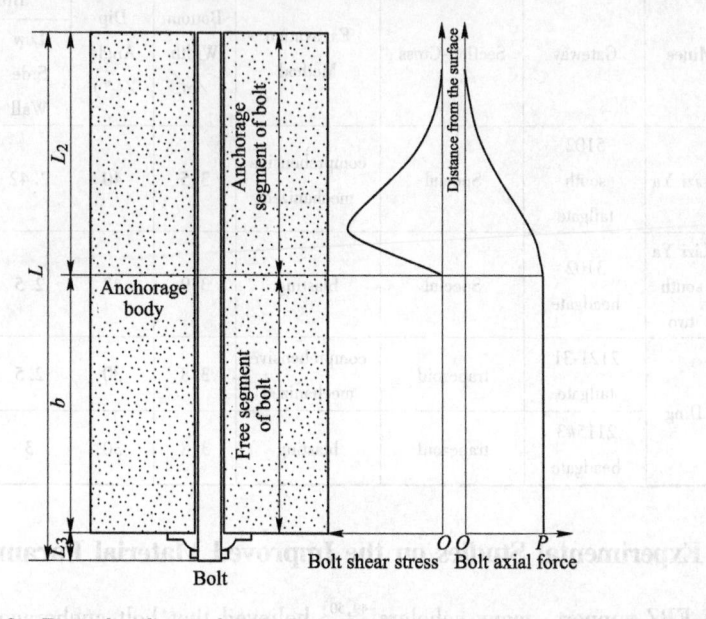

Figure 7.9　Force distribution diagrammatic sketch of an anchorage body and bolt

There are three kinds of anchorage methods in terms of the anchorage length[40]:

(1) When the anchorage length is not greater than 500mm or 1/3 of the borehole length for the bolt, it is called end anchorage.

(2) When the anchorage length is not less than 90% of the borehole length for the bolt, it is called full length anchorage.

(3) When the anchorage length is between the length of an end anchorage and a full length anchorage, it is called extensible anchorage. Further analysis reveals that:

1) The dispersion range of pretension force of a bolt with full length anchorage is minimum relative to the other two anchorage methods. In addition, the jumbolter often stop running during stiring bolt with resin cartridge when the bolt was using full length anchorage.

2) The bolt with end anchorage can increase the dispersion range of pretension force, but the anchorage length is shorter, resulting in low safety coefficiency to a bolt in a large EBZ, and the bolt will lose efficacy easily under an action of high pretension force.

Therefore, based on the above analysis, the extensible anchorage method was determined to best support the gateway surrounding rock with a large EBZ.

7.7.2 The Improved Mechanical Performance of a Bolt

Anchorage force is a useful indicator to assess the support system, and is dependent upon the complicated geological engineering factors and loading conditions to which it is subjected[55]. The strength parameters of the bolt itself, the bond force at the interface between the bolts and the rock, and the surrounding rock property, as well as the bearing capacity of the load point on the surface are all primary factors found in previous studies. Field results, using high-performance steel in the rock bolt, show the load bearing capacity on a surface is reliable, and will be not discussed in this paper. Hence, this section will analyze whether the current industrial technology can provide high pretension force under the mentioned another three primary factors.

7.7.2.1 Bolt Pullout Test

For the purpose of investigating the mechanical properties of a bolt used in an extensible anchorage method, the strength parameters of the bolt itself should first be tested. Hence, four hot-rolled fine thread resin bolts with a diameter of 22mm and a length of 2400mm were static pull tested. The results are listed in Table 7.6, and the bolt fracture surface was shown in Figure 7.10.

Table 7.6 Testing results of the mechanical properties of bolts

Number	Yield load /kN	Yield strength /MPa	Breaking load /kN	Breaking strength /MPa	Percentage elongation/%
1	255	671	306	805	18.2
2	250	658	303	797	19.5
3	254	668	304	800	Over gauge length
4	249	655	300	789	Over gauge length
Average value	252	663	304	798	18.8

The average values of yield load, breaking load and percentage elongation are 252 kN, 304kN and 18.8% respectively, indicating the strength parameters of this type of bolt have significantly increased compared to many previously used bolts with a tensile breaking strength of 100kN to 150kN in coal mines[41,55]. They are characterized by their high load capacity.

· 128 · 7 The Failure Characteristics and the Supporting Technology for Pre-Retaining Gateway

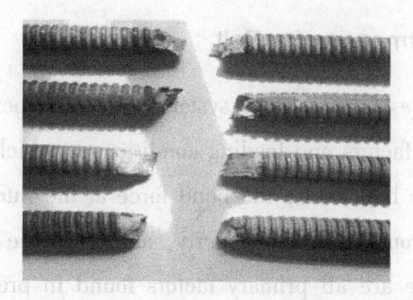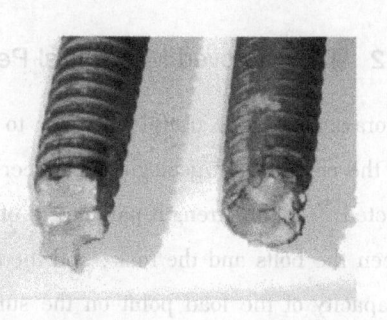

Figure 7.10　Fracture surface of tensile broke bolts

7.7.2.2　Bond Force Test in a Laboratory

In addition to the strength parameters of the bolt itself, the bond force of the grouted part is also a useful factor to reflect the anchorage force. Two kinds of anchorage lengths of bolts mentioned above were laboratory tested to investigate the capability of providing suitable bond force. The results are listed in Table 7.7, and parts of the cross-section of anchored segment are shown in Figure 7.11.

Table 7.7　The results of laboratory bond force tests for bolt pullout

Anchorage length/mm	NO.	Cylinder pressure/MPa	Anchorage force/kN	Failure mode	Comment
330	1	47.85	271.4	Sliding failure	Resin cartridge: MKφ2835; Test steel tube inner diameter of 30mm
	2	42.69	242.1	Sliding failure	
	3	41.41	234.9	Sliding failure	
125	1	18.63	105.7	Sliding failure	
	2	17.52	99.4	Sliding failure	
	3	3.57	20.2	Sliding failure	
	4	15.79	89.6	Sliding failure	

It observed that sliding failure occurred at the bolt interface of a tube. The average anchorage force of a bolt with an anchorage length of 300mm, could reach up to 249.5 kN, almost equal to the tensile yield load of the bolt. This is much higher than that of a bolt with an anchorage length of 125mm. Therefore, it is reasonable to believe that when the anchorage length is longer than 300mm of extensible anchorage, the anchorage force will increase. It also reflects that the resin cartridge of MK φ2835 has superior performance and can provide a higher bond force.

In addition, the hot-rolled fine thread resin bolt not only has a higher anchorage force in a laboratory, but also has a self-locking effect. It can be seen that the matching nut

7.7 Experimental Studies on the Improved Material Parameters of Bolts · 129 ·

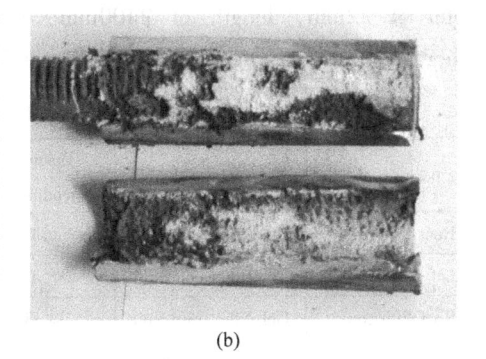

(a)   (b)

Figure 7.11   Cross-section of anchored segment

(a) anchorage length of 300mm; (b) Anchorage length of 125mm

did not retreat after it locked the bolt when the loading force reached 300kN in the testing process, as shown in Figure 7.12.

Figure 7.12   The nut status when the bolt load was 300kN

7.7.2.3   Anchorage Force Tests in the Field

Laboratory tests on bolts can represent the behavior of bolts when applying a load at one end of the bolt. However, they cannot reproduce the field conditions and the actual rock mass state.

Field tests can provide quantitative data to examine the performance of rock mass supports. Hence, anchorage force pullout experiments were conducted in an immediate roof of a gateway in the field. The lithology of the immediate roof at the test site was poor, containing sandy mudstone and coal streaks, with a thickness of 3.32m, and an uniaxial compressive strength of less than 20MPa. The dimensions of the bolts tested were: diameter of 22mm, length of 2400mm, anchorage length of 1500mm (extensible anchorage). The results are listed in Table 7.8.

Table 7.8 The pullout results of bolts anchorage force in field tests

Test position of bolt	Near gob/coal wall	Distance from heading end/m	Anchorage force/kN	Comment
Roof first row	Gob/the second bolt	12.8	88	Not pullout enough
Roof second row	Gob/the second bolt	12.2	132	Sleeve is snapped
Roof fifth row	Gob/ the third bolt	8.8	88	Not pullout enough
Roof sixth row	Coal wall/ the second bolt	8.0	88	Not pullout enough

The highest anchorage force was 132kN when the sleeve snapped, parts of the sleeve in the field tests are shown in Figure 7.13. On the other hand, it is reasonable to expect that the poor lithology could provide a higher anchorage force than 132kN in the field when experiencing a bolt breaking or a bolt sliding failure.

Figure 7.13 The sleeve used in a pullout test in the field

7.7.2.4 Method for Increasing Pretension Force

The above studies revealed that the bolt, resin cartridge and poor lithology have the ability to bear high pretension force. But, it is a question of how to conveniently increase the pretension force in the field. Thus, engineers invented a torque amplifier, which can make the bolt's original torque increase 3~3.5 times, greatly improving the bolt pretension force[57]. Moreover, the torque size can be tested by digital display torque wrench. The torque amplifier and digital display torque wrench using in field were shown in Figure 7.14.

(a)   (b)

Figure 7. 14   Torque amplifier and torque wrench used in the field

(a) torque amplifier; (b) torque wrench

Therefore, it can be seen that high pretension force can be provided in the field to support the surrounding rock in a large EBZ by using the improved bolt and extensible anchorage method.

7. 8   Support Effect Analyses of Bolt and Surround Rock

7. 8. 1   Support Theory Assessment of Gateways Broken Zones

For large broken widths, bolts installed within an EBZ can form a combined arch structure in the surrounding rocks. The support design should be adjusted to utilize the self-bearing capacity of the combined arch, and the support parameters design mainly depends on the combined arch theory[52]. Previous research (in 1994) considered the bolt support parameter calculation diagram of a combined arch as shown in Figure 7. 15, and the thickness of a combined arch could be determined by:

$$b = \frac{L\tan\beta - a}{\tan\beta} \tag{7.5}$$

Where b——the thickness of a combined arch;

   L——the bolt's effective length;

   β——the bolt control angle in an EBZ, $\beta = 45°$;

   a——the inter-row spacing of bolts.

To improve the bearing capacity of a combined arch in a large EBZ, it is necessary to increase the strength and thickness of the combined arch. Further study proposed the thickness of a combined arch is the basic parameter to calculate the bearing capacity of a combined arch, and with an increase in the thickness of a combined arch at a certain range, its bearing capacity would increase[58].

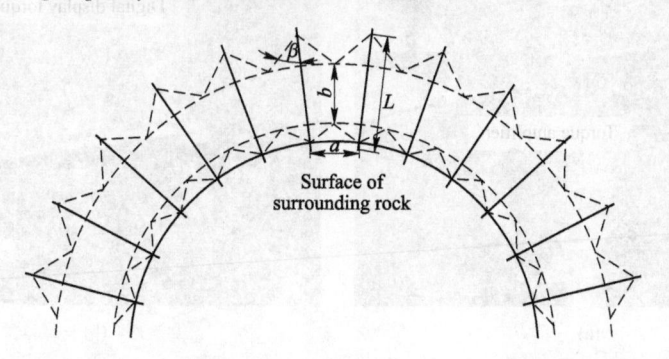

Figure 7.15 Parameter calculation diagram of combined arch

Itwasnoted that the support parameter-calculation method in Equation (7.5) was proposed under the condition that a bolt with low tensile strength, low anchoring force, and low pretension force, usually reflecting a passive support. Afterwards, with the invention of the high tensile strength resin bolt and steel and mesh, and the development of active support technology for pretension force bolt, the parameter selection of Equation(7.5) should be discussed as follows:

(1) Under the action of bolt pretension force and high strength steel to its dispersion on the surface of the surrounding rock, whether the control angle of β is still 45°.

(2) The bolt usually was full-length anchored, whether it effectively increased the thickness of a combined arch, and formed a self-bearing structure.

(3) The thickness of a combined arch b was usually determined by an experience of 0.85m, 0.9m, 1m, 1.2m with different sizes.

Therefore, aiming at the above problems, it is very necessary to conduct further analyses on the combined arch support theory for a large EBZ according to modern bolt mechanical properties and numerical simulation technologies.

7.8.2 Stress Diffusion Analysis for Anchored Rock Mass

In general, bolts and steel are used together. Thus, high pretension force bolt, high strength steel and extensible anchorage method were used to study the improving effect of a combined arch in a large EBZ, and the finite differential method of a FLAC3D was employed as the analytical tool.

To reach a deeper understanding of the diffused characteristics of pretension force in surrounding rock when the rock bolt pretension force increases, pretension forces of

30kN, 60kN, 90kN and 120kN were used and calculated. Moreover, the numerical model utilized the Mohr-Coulomb failure criterion, and one sandstone material was endowed with the physical and mechanical parameters: elastic modulus of 30GP, Poisson's ratio of 0. 3, a density of 2400kg/m³, a cohesion of 10MPa, a friction Angle of 30°, and tensile strength of 2MPa. Simultaneously, the steel was endowed with elastic material, with an elastic modulus of 100GP and a Poisson's ratio of 0. 25. In addition, two bolts, with a diameter of 22mm, a length of 2400mm and an inter space of 800mm, were used for the study. It must be emphasized that the anchorage length of 1000mm was used and the bolt end was anchored with 10mm to simulate bolt plate.

Furthermore, the rock bolts were represented as built-in 'cable' elements. For resin-grouted rock bolts, a stiffness of 2×10^{10}N/m, a cohesive strength of 4×10^{8}N/m, a friction angle of 33°, a cross-sectional area of 3. 8×10^{-4}m², an elastic modulus of 205GPa, a tensile yield strength of 250kN were assigned to the 'cable' element in this study. The calculation results are as shown in Figure 7. 16.

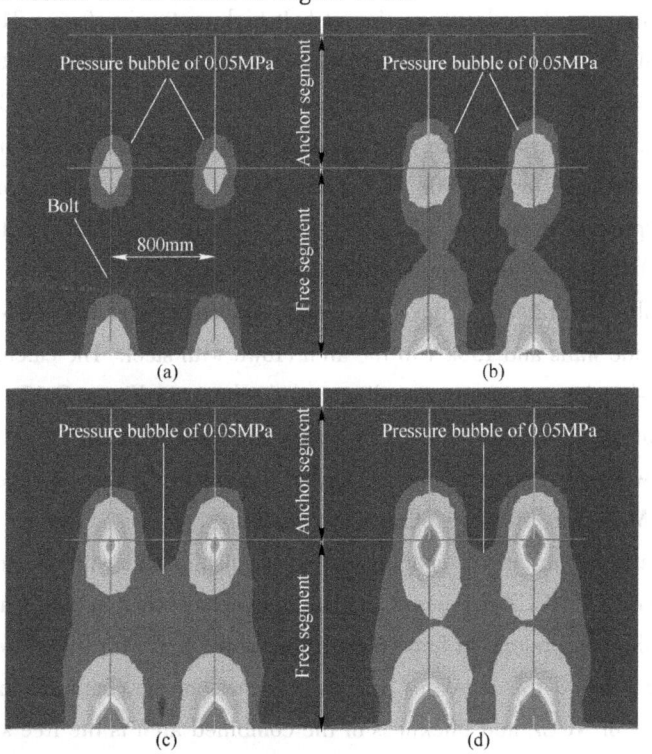

Figure 7. 16   Stress distribution contours in surrounding rock
(a) Pretension force of 30kN; (b) Pretension force of 60kN;
(c) Pretension force of 90kN; (d) Pretension force of 120kN

It is said that when the compressive stress value is greater than 0.05MPa in rock mass, the rock mass is effectively controlled by bolt anchorage force[4]. Figure 7.16 schematically shows that:

(1) With the increase of pretension force, the range of pressure bubble of 0.05MPa increases. First, the pressure bubbles began to connect in free segments in one bolt when the pretension force was 60kN. Second, the pressure bubbles connected between bolts when the pretension force was 90kN. Finally, the connected area had no further increase when the pretension force is 120kN.

(2) Under the action of steel and pretension force, the pressure bubble almost distributed within the rock mass of the bolt free segment, illustrating that the initial extrusion reinforcement range was in the rock mass of the bolt free segment.

7.8.3 Initial Load-Bearing Zone of Gateway Surrounding Rock

The initial distribution area of the combined arch embodied the load-bearing capacity in a gateway surrounding rock under the action of bolt high pretension force. This section will study the initial load-bearing zone, and a gateway with an arched cross section as the calculation example, and with the above numerical model parameters of FLAC3D as the material. The following statistics were documented: (1) bolt inter-row spacing of 800mm× 800mm, bolt pretension force of 90kN in two side walls and 120kN in the arch crown with steel, and bolt pretension force of 120kN in two side walls and 150kN in the arch crown with steel; (2) bolt inter-row spacing of 700mm×700mm, bolt pretension force of 90kN in two side walls and 120kN in the arch crown with steel, and bolt pretension force of 120kN in two side walls and 150kN in the arch crown with steel. The calculated results of the initial load-bearing zone in surrounding rock is shown in Figure 7.17.

It is necessary to define the concept of a non-uniform compressed zone (NUCZ) that distributes between the anchor segment end and the combined arch surface, as shown in Figure 7.12. Moreover, it shows other information:

(1) When the bolt inter-row spacing remains constant and with the increase of bolt pretension force, the NUCZ decreases and the thickness of the combined arch increases.

(2) When the bolt pretension force remains constant, and with the decrease of bolt inter-row spacing, the NUCZ decreases and the thickness of the combined arch increases.

(3) The sum of NUCZ and thickness of the combined arch is the free segment range, and with the increase of bolt pretension force and the decrease of bolt inter-row spacing, the NUCZ is close to zero, presenting that the free segment of anchorage body is the initial load-bearing zone (combined arch area).

7. 8   Support Effect Analyses of Bolt and Surround Rock   · 135 ·

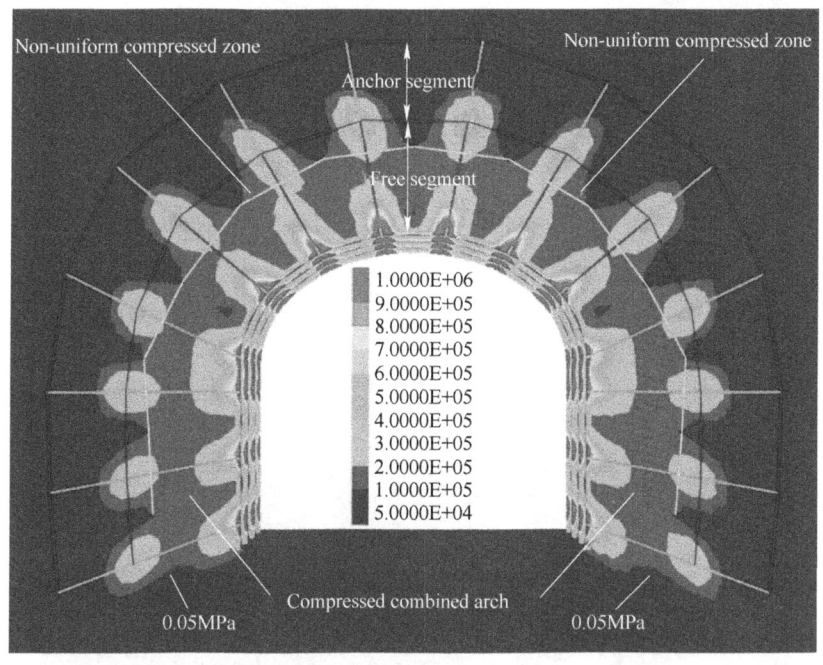

(a)

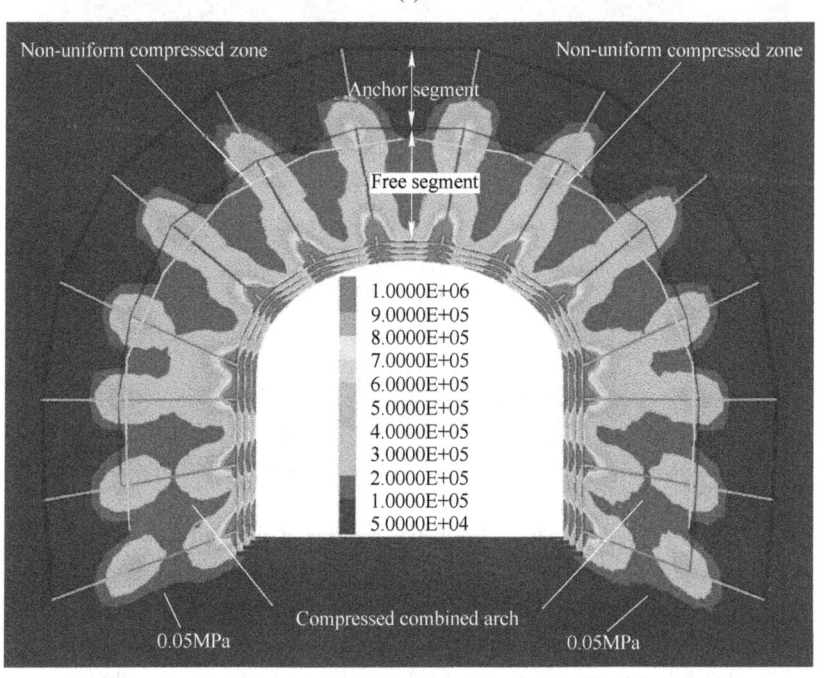

(b)

· 136 · 7　The Failure Characteristics and the Supporting Technology for Pre-Retaining Gateway

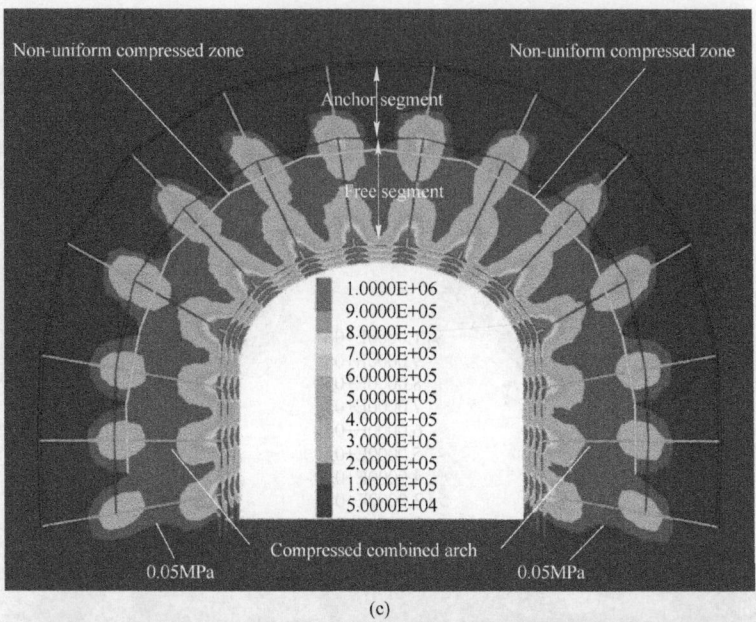

(c)

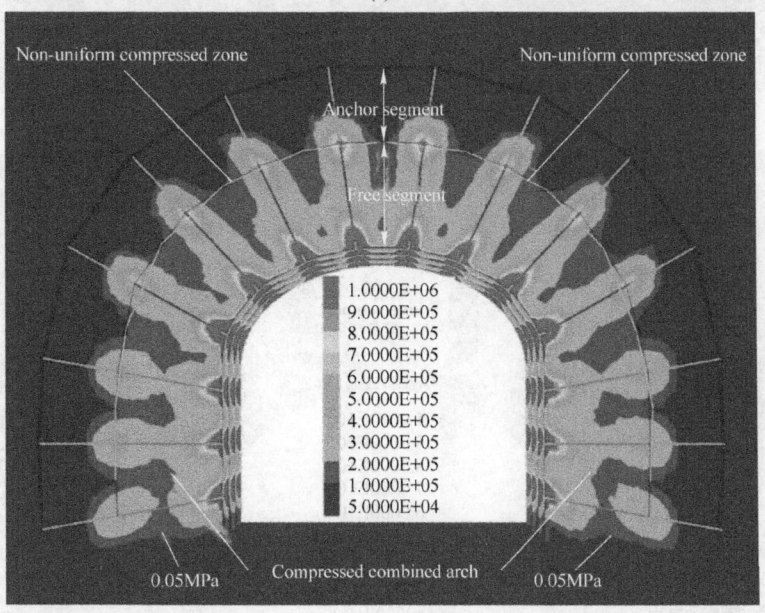

(d)

Figure 7.17　Total stress distribution contours of anchored rock under
different pre-tensioned force with steel belt

(a) Bolts with inter-row spacing of 800mm and pretension force of 90~120kN;
(b) Bolts with inter-row spacing of 800mm and pretension force of 120~150kN;
(c) Bolts with inter-row spacing of 700mm and pretension force of 90-120kN;
(d) Bolts with inter-row spacing of 700mm and pretension force of 120~150kN

The thickness of the initial load-bearing zone of gateway surrounding rock is the length of bolt free segment, as shown in Figure 7.17. Thus, the thickness of combined arch b is equal to the bolt full length L_1 minus the bolt anchorage length L_2 and the bolt end exposed length L_3, as shown in $b = L_1 - L_2 - L_3$.

Furthermore, it can reference[58] to calculate the initial bearing capacity of a combined arch in a large EBZ. Therefore, the bolt with high pretension force will provide high bearing capacity to surrounding rock, and prevent large deformation and failure at the beginning of the bolt installation. It has remarkable advantages relative to the bolt with no steel and low bolt pretension force in surrounding rock.

7.9 Supporting Technology

The above investigation demonstrated that the coal seam dip angle mostly varies from $25° \sim 40°$ in coal mines; this finding indicates that most of the gateways would present an asymmetric failure type to the gateway centre. In previous studies, owing to the asymmetric coal and rock mass structure and stress evolution, bedding-plane shear slip and high stress dilatancy mechanism caused asymmetric deformation failure patterns in the gateways[36,37]. For the broken coal and rock mass within the EBZ, a large hulking force would be produced, become larger with increasing broken width, and become the main supporting object[1].

However, for a field support system, the bolts and cables usually have a low pretension force, and the support parameters were basically consistent in the two side walls and the roof. Moreover, the two side walls bolt lengths were equal for most parts of the coal mine, presenting a phenomenon of a symmetrical support method supporting asymmetrical failure structure, thus leading to multiple occurrences of large deformations and slice in the surrounding rock.

Therefore, an asymmetric control technique was required to support the gateways with the EBZ characteristics above using bolts, cables and shotcrete combined with steel mesh and steel belts. In reality, the bolts installed within the EBZ can form a combined arch structure in the surrounding rocks[4]. Because of the larger broken width in roof and high side wall, the improved support parameters could lead to the arch crown, and the arch springing thickness in the roof and high side wall were larger than that in the low side wall ($L_a > L_b$), enhancing the bearing capacity of the larger failure zone, as shown in Figure 7.18. This asymmetric control technique was conducive to maintaining the stability of the gateway. At present, asymmetric control technology has achieved the expected effect in many soft fractured rock mass roadways, especially in deep mine roadways[38,39].

· 138 · 7 The Failure Characteristics and the Supporting Technology for Pre-Retaining Gateway

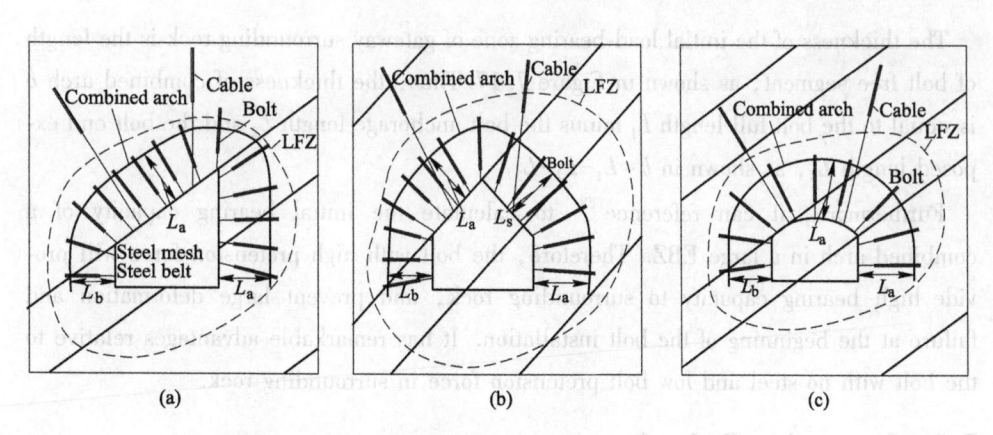

Figure 7.18 The cross-section of the support system

(a) Support for trapezoid cross-section; (b) Support for special cross-section;

(c) Support for arch or inclined arch cross-section

In the field, engineers can determine the detailed supporting parameters according to the geological conditions. When dip angle of the coal seam is larger, high strength bolt can be used, the bolt lengths and bolt anchorage lengths in roof and high side wall can be increased, and the pretension force of bolts and cables can be improved.

This support system used an extensible bolting method, and the anchorage length was not less than 50% of the bolt length to increase the anchoring force in larger EBZ. This support system could increase the sphere of pretension force, and the effect of pretension force anchor in soft rock is better than that in hard rock[40]. Cable reinforcement must be added to the support system with an overhanging effect and anchored beyond the EBZ in the stable roof rocks[4].

The analysis above provided reference for support design.

7.10 Chapter Summary

GPR and 3DEC models were employed to determine the state of the EBZ in the surrounding rock of 55 gateways at coal mines in Sichuan Province, China. The primary conclusions with regard to gateways with similar geological conditions were as follows:

(1) It was found that the minimum value was 1.5m, and the maximum value was 3.5m, and the EBZ was large. In general, the broken widths in the roof and high side wall were relatively large compared to those in the low side wall. It was found that the broken width was significantly influenced by the buried depth of the gateway, coal seam dip angle, coal seam thickness, and immediate roof thickness. Larger coal seam, immediate roof thickness, and smaller coal seam dip angle had relatively larger broken widths

in the roof and high side wall. Furthermore, the broken widths also increased with the increase of the buried depth. However, the change in trend was not evident. In reality, broken width is the result of a combined action, and as a multi-factor single index for evaluating the damage degree of surrounding rock, this phenomenon sufficiently demonstrated the practicability and scientificity of the evaluate index.

(2) Combined with the tested broken widths and plastic zones in the 3DEC model, each gateway EBZ cross-section diagram was mapped with a smooth curve. It was found that the EBZ distribution shapes were basically elliptical. The long axis was along the seam inclination direction, the short axis was along the vertical direction of the rock layer, and the failure extent was greater in the seam inclination direction than that in the vertical direction of the rock layer in the surrounding coal and rock mass. Also, the elliptical shape had little relationship with the gateway cross-section shape, presenting asymmetric failure formation.

(3) To make a further analysis of the combined arch support theory for large EBZ according to modern bolt mechanical properties and numerical simulation technology, we conducted experimental investigations in a laboratory and in the field. It found that the hot-rolled fine thread resin bolt had high tensile strength presenting superior mechanical properties. Moreover, the pullout experiment illustrated that the poor lithology can provide high anchorage force with a bolt extensible anchorage method. Finally, the invented torque amplifier can greatly improve bolt pretension force, and meet high pretension force support requirement.

(4) In addition, after a FLAC3D numerical simulation analysis, it found that the pretension force main diffusion sphere was in the free segment zone of the surrounding rock. Further analysis showed that the initial load-bearing zone thickness of a combined arch structure in a large EBZ can be expressed by the bolt free segment length, when using high mechanical property bolts and steel with high pretension force, and finally, clearly put forward the bolt length selection rule based on the combined arch support theory.

(5) An asymmetry control technique can be used for gateway stabilization. The bolt length in the roof and the high side wall should be larger than that in the low side wall, and the reinforcement cable must be anchored in the stable surrounding rocks. The bolts within the EBZ formed a combined arch structure in the surrounding rocks, and the thickness of the arch crown and arch springing should be larger in the roof and high side wall, respectively. In addition, cables should have an overhanging and reinforcing effect for the combined arch. This support method only provided a reference for field support design.

References

[1] Dong F, Song H, Guo Z, et al. Roadway support theory based on broken rock zone[J]. China Coal Soc, 1994, 19:21-32.

[2] Pusch R, Stanfors R. The zone of disturbance around blasted tunnels at depth[J]. Rock Mech Min Sci, Geomech. Abstr, 1992, 29:447-456.

[3] Kelsall P C, Case J B, Chabannes C R. Evaluation of excavation-induced changes in rock permeability[J]. Rock Mech Min Sci, Geomech Abstr, 1984, 21:123-135.

[4] Wang H, Jiang Y, Xue S, et al. Assessment of excavation damaged zone around roadways under dynamic pressure induced by an active mining process[J]. Rock Mec Min Sci, 2015, 77: 265-277.

[5] Renaud V, Balland C, Verdel T. Numerical simulation and development of data inversion in borehole ultrasonic imaging[J]. Appl Geophys, 2011, 73:357-367.

[6] Tsang C F, Bernier F, Davies C. Geohydromechanical processes in the excavation damaged zone in crystalline rock, rock salt, and indurated and plastic clays——In the context of radioactive waste disposal[J]. Rock Mech Min Sci, 2005, 42:109-125.

[7] Sato T, Kikuchi T, Sugihara K. In-situ experiments on an excavation disturbed zone induced by mechanical excavation in neogene sedimentary rock at tono mine, central Japan[J]. Eng Geol, 2000, 56:97-108.

[8] Schuster K, Alheid H J, Böddener, D. Seismic investigation of the excavation damaged zone in opalinus clay[J]. Eng Geol, 2001, 61:189-197.

[9] Malmgren L, Saiang D, Töyrä J, et al. The excavation disturbed zone (EDZ) at Kiirunavaara mine, sweden——By seismic measurements[J]. Appl Geophys, 2007, 61:1-15.

[10] Li S, Feng X T, Li Z, et al. Evolution of fractures in the excavation damaged zone of a deeply buried tunnel during TBM construction[J]. Rock Mech Min Sci, 2012, 55:125-138.

[11] Tan Y L, Yu F H, Chen L. A new approach for predicting bedding separation of roof strata in underground coalmines[J]. Rock Mech Min Sci, 2013, 61:183-188.

[12] Homand-Etienne F, Hoxha D, Shao J F. A continuum damage constitutive law for brittle rocks. Comput[J]. Geotech, 1998, 22:135-151.

[13] Golshani A, Oda M, Okui Y, et al. Numerical simulation of the excavation damaged zone around an opening in brittle rock[J]. Rock Mech Min Sci, 2007, 44:835-845.

[14] Li L C, Liu H H. A numerical study of the mechanical response to excavation and ventilation around tunnels in clay rocks[J]. Rock Mec. Min. Sci. 2013, 59, 22-32.

[15] Pellet F, Roosefid M, Deleruyelle F. On the 3D numerical modelling of the time-dependent development of the damage zone around underground galleries during and after excavation[J]. Tunn Undergr Space Technol, 2009, 24:665-674.

[16] Gao F Q, Stead D. The application of a modified voronoi logic to brittle fracture modelling at the laboratory and field scale[J]. Rock Mech Min Sci, 2014, 68:1-14.

[17] Kang H P, Lin J, Fan M J. Investigation on support pattern of a coal mine roadway within soft

rocks——A case study[J]. Coal Geol, 2015, 140:31-40.

[18] Pérez-Gracia V, García García F, Rodriguez Abad, I. Gpr evaluation of the damage found in the reinforced concrete base of a block of flats: A case study[J]. NDT E Int, 2008, 41:341-353.

[19] Xiang L, Zhou H, Shu Z, et al. GPR evaluation of the damaoshan highway tunnel: A case study [J]. NDT E Int, 2013, 59:68-76.

[20] McCann D M, Forde M C. Review of ndt methods in the assessment of concrete and masonry structures[J]. NDT E Int, 2001, 34:71-84.

[21] Orbán Z, Gutermann M. Assessment of masonry arch railway bridges using non-destructive in-situ testing methods[J]. Eng Struct, 2009, 31:2287-2298.

[22] Benedetto A, Pensa S. Indirect diagnosis of pavement structural damages using surface GPR reflection techniques[J]. Appl Geophys, 2007, 62:107-123.

[23] Porsani J L, Sauck W A, Júnior A O S. GPR for mapping fractures and as a guide for the extraction of ornamental granite from a quarry: A case study from southern brazil[J]. Appl Geophys, 2006, 58:177-187.

[24] Shaw M R, Millard S G, Molyneaux T C K, et al. Location of steel reinforcement in concrete using ground penetrating radar and neural networks[J]. NDT E Int, 2005, 38:203-212.

[25] Loizos A, Plati C. Accuracy of pavement thicknesses estimation using different ground penetrating radar analysis approaches[J]. NDT E Int, 2007, 40:147-157.

[26] Church R H, Webb W E, Boyle J R. Ground-penetrating radar for strata control, report of investigations[M]. Washington, DC, USA: united states bureau of mines, 1985.

[27] Zhang P, Li Y, Zhao Y, Guo L. Application and analysis on structure exploration of coal seam by mine ground penetrating radar[C]. In Proceedings of the 2012 14th International Conference on Ground Penetrating Radar, 2012,6(4-8):469-472.

[28] Strange A D, Ralston J C, Chandran V. Application of ground penetrating radar technology for near-surface interface determination in coal mining[C]. In Proceedings of the 2005 IEEE International Conference on Acoustics, Speech, and Signal Processing, 2005,3(18-23):701-704.

[29] Koarolu S, Erik N Y. Ground penetrating radar (GPR) method of geological properties of coal seams near the surface[C]. In Proceedings of the 14th International Multidisciplinary Scientific Geoconference and EXPO, 2014,6(17-26):467-474.

[30] Bai B, Zhou J. Advances and applications of ground penetrating radar measuring technology. Chin[J]. Rock Mech Eng, 2001, 20:527-531.

[31] Song H, Wang C, Jia Y. Principle of measuring broken rock zone around underground roadway with gpr and its application[J]. China Univ Min Technol, 2002, 31:370-373.

[32] Gu J, Shen J, Chen A, et al. Model testing study of strain distribution regularity in rock mass caused by prestressed anchorage cable[J]. Rock Mech Eng, 2000, S1:917-921.

[33] Kang H P, Jiang P F, Cai, J. F. Test and analysis on stress fields caused by rock bolting[J]. China Coal Soc, 2014, 39(8):1521-1529.

[34] Lin J, Shi Y, Sun Z, et al. Large scale model test on the distribution characteristics of the prestressed field of end-anchored bolts[J]. Rock Mech Eng, 2016, 35(11):2237-2247.

[35] Guo X, Mao X, Ma C, et al. Bolt support mechanism based on elastic theory[J]. Min Sci Technol, 2013, 23(4):469-474.

[36] Showkati A, Maarefvand P, Hassani H. Stresses induced by post-tensioned anchor in jointed rock mass[J]. Cent South Uni, 2015, 22(4):1463-1476.

[37] Ranjbarnia M, Fahimifar A, Oreste P. A simplified model to study the behavior of pre-tensioned fully grouted bolts around tunnels and to analyze the more important influencing parameters[J]. Min Sci, 2014, 50(3):533-548.

[38] Ding X, Sheng Q, Han J, et al. Numerical simulation testing study on reinforcement mechanism of prestressed anchorage cable[J]. Rock Mech Eng, 2002, 21(7):980-988.

[39] Wang J, Kang H, Gao F. Numerical simulation study on load transfer mechanisms and stress distribution characteristics of cable bolts[J]. China Coal Soc, 2008, 33(1):1-6.

[40] Wei S J, Li B F. Anchor bolt body formation and instability mode under the influence of anchoring pretension[J]. China Coal Soc, 2013, 38(12):2126-2132.

[41] Kang H. Sixty years development and prospects of rock bolting technology for underground coal mine roadways in China[J]. China U Min Techno, 2016, 45(6):1071-1081.

[42] Itasca. 3DEC——3 dimensional distinct element code[D]. Minneapolis Itasca Consulting Group Inc, 2013.

[43] Gao F, Stead D, Kang H, et al. Discrete element modelling of deformation and damage of a roadway driven along an unstable goaf——A case study[J]. Coal Geol, 2014, 127:100-110.

[44] Davis J L, Annan A P. Ground-penetrating radar for high-resolution mapping of soil and rock stratigraphy. Geophys[J]. Prospect, 1989, 37:531-551.

[45] Xie J L, Xu J L. Ground penetrating radar-based experimental simulation and signal interpretation on roadway roof separation detection[J]. Arab J Geosci, 2015, 8:1273-1280.

[46] Fan K G, Jiang J Q. Deformation failure and non-harmonious control mechanism of surrounding rocks of roadways with weak structures[J]. China Univ Min Technol, 2007, 36:54-59.

[47] Cao S G, Zou D J, Bai Y J, et al. Surrounding rock control of mining roadway in the thin coal seam group with short distance and "three soft"[J]. Min Saf Eng, 2011, 28:524-529.

[48] Sun X, Zhang G, Cai F, et al. Asymmetric deformation mechanism within inclined rock strata induced by excavation in deep roadway and its controlling countermeasures[J]. Rock Mech Eng, 2009, 28:1137-1143.

[49] Yu Y, Bai J, Wang X, et al. Study on asymmetric distortion and failure characteristics and stability control of soft rock roadway[J]. Min Saf Eng, 2014, 31:340-346.

[50] Zheng X G, Zhang N, Xue F. Study on stress distribution law in anchoring section of prestressed bolt[J]. Min Saf Eng, 2012, 29:365-370.

[51] Xiao J, Feng X, Lin D. Influence of blasting round on excavation damaged zone of surrounding rock[J]. Rock Mech Eng, 2010, 29:2248-2255.

[52] Wang H, Jiang Y, Xue S, et al. Assessment of excavation damaged zone around roadways under dynamic pressure induced by an active mining process[J]. Rock Mech Min Sci, 2015, 77:265-277.

References

[53] Zhou H, Xu R C, Zhang C Q, et al. Research on effect of interior bonding section length of pre-stressed anchor rod[J]. Rock Soil Mech, 2015, 36(9): 2688-2694.

[54] Ma S, Nemcik J, Aziz N. An analytical model of fully grouted rock bolts subjected to tensile load [J]. Constr Build Mater, 2013, 49:519-526.

[55] Nemcik J, Ma S, Aziz N, et al. Numerical modelling of failure propagation in fully grouted rock bolts subjected to tensile load[J]. Rock Mech Min Sci, 2014, 71:293-300.

[56] Li C C, Stjern G, Myrvang A. A review on the performance of conventional and energy-absorbing rockbolts[J]. Rock Mech Geotech Eng, 2014, 6(4):315-327.

[57] Zou D, Li M, Gong L, et al. Study on the rib spalling and the treatment during driving in strong outburst and medium thick coal seam[J]. Coal Min Mod, 2015, 4:81-83.

[58] Yu W, Gao Q, Zhu C. Study of strength theory and application of overlap arch bearing body for deep soft surrounding rock[J]. Rock Mech Eng, 2010, 29(10):2134-2142.

Attachment　Evaluation Results of Adaptability of Gob-side Entry Retaining

Table 1　Adaptability evaluation results of gob-side entry retaning for PanMei company

Gateway No.	Name of coal seams	Buried depth/m	Dip angle /(°)	Thickness /m	Lithology (f)	TICIR/ N	Roof integrity	Adaptive grades
1	38	317	15	1. 7	7	3. 6	70	III
2	39-1	315	23	3. 2	6	0. 8	55	III
3	39-1	300	23	3. 2	6	0. 8	55	IV
4	39-1	361	18	2. 3	5	1. 8	55	IV
5	39-1	409	18	3. 4	5	1. 2	60	IV
6	38	418	27	1. 6	6	1. 6	70	IV
7	38	401	16	1. 6	6	1. 6	70	III
8	39-1	174	23	2. 9	5	1. 1	55	IV
9	21-3	321	31	1. 7	8	10. 6	70	I
10	15	468	34	1. 3	4	10. 8	65	III
11	24-1	381	30	0. 9	4	3. 9	75	II
12	18	227	23	0. 9	4	4. 4	70	II
13	4	326	38	2. 8	6	1. 1	65	V
14	21	393	28	2. 0	6	2. 2	70	III
15	21	478	28	2. 0	6	2. 2	70	IV
16	23	524	26	2. 8	4	2. 0	75	IV
17	27	653	29	1. 4	4	0. 9	70	II
18	24	350	37	2. 7	6	0. 4	75	V
19	24	400	36	2. 0	6	1. 6	75	II
20	23	275	36	2. 0	6	1. 6	75	II
21	18	320	61	3. 5	6	2. 7	70	II
22	18	320	61	3. 5	6	2. 7	70	II
23	18	326	61	3. 1	6	3. 0	70	II
24	3	280	69	2. 2	6	0. 4	75	I
25	15	300	37	5. 0	6	0. 8	65	V
26	15	300	37	5. 0	6	0. 8	65	V
27	23	337	36	2. 0	6	1. 6	75	II
28	23	337	37	2. 0	6	1. 6	75	II
29	24	350	37	2. 7	6	0. 4	75	V

Adaptability evaluation results of gob-side entry retaning for PanMei company.

Attachment   Evaluationresults of adaptability of gob-side entry retaining   · 145 ·

Table 2   Adaptability evaluation results of gob-side entry retaning for GuangNeng company

Gateway No.	Name of coal seams	Buried depth/m	Dip angle /(°)	Thickness /m	Lithology (f)	TICIR /N	Roof integrity	Adaptive grades
1	K1	480	36	2. 5	6	3. 4	85	V
2	K1	480	36	2. 5	6	3. 4	85	V
3	K1	550	35	2. 4	4	3. 0	70	Ⅲ
4	K2-1	372	18	3. 2	4	1. 2	53	Ⅳ
5	K2-1	372	18	3. 2	4	1. 2	54	Ⅳ
6	K1	600	40	2. 2	6	2. 3	70	V
7	K2-1	530	18	1. 3	4	9. 3	52	Ⅲ
8	K2-1	350	12	1. 4	5	4. 4	61	Ⅱ
9	K1	550	7	1. 7	6	9. 6	85	I
10	K1	490	7	1. 7	6	9. 6	86	I
11	K1	430	54	2. 1	6	2. 1	60	Ⅲ
12	K1	333	49	2. 0	6	2. 3	70	Ⅲ
13	K1	425	49	1. 9	6	2. 4	70	Ⅲ
14	K1	515	49	1. 9	6	2. 4	70	Ⅲ
15	K1	445	47	2. 4	5	6. 1	55	Ⅳ
16	K1	500	48	2. 2	5	6. 9	55	Ⅳ
17	K1	500	48	2. 2	5	6. 9	55	Ⅳ
18	K1	496. 5	27	0. 9	6	6. 4	48	Ⅳ
19	K1	424. 5	27	0. 9	6	6. 4	48	Ⅳ

Adaptability evaluation results of gob-side entry retaning for GuangNeng company.

Table 3   Adaptability evaluation results of gob-side entry retaning for FuRong company

Gateway No.	Name of coal seams	Buried depth/m	Dip angle /(°)	Thickness /m	Lithology (f)	TICIR /N	Roof integrity	Adaptive grades
1	B2	345	8	1. 3	5	1. 5	85	Ⅱ
2	B3	448	11	1. 2	4	1. 5	75	Ⅱ
3	C1	375	12	1. 3	7	1. 2	85	I
4	B2	414	13	1. 3	6	0. 2	75	Ⅱ
5	C1	420	9	0. 8	7	0. 8	85	I
6	B4	403	23	1. 4	5	1. 6	55	Ⅳ
7	B4	395	24	1. 5	5	1. 4	45	Ⅳ
8	C1	338	24	1. 2	5	1. 5	55	Ⅳ

· 146 ·   Attachment Evaluationresults of adaptability of gob-side entry retaining

Continued Table 3

Gateway No.	Name of coal seams	Buried depth/m	Dip angle /(°)	Thickness /m	Lithology (f)	TICIR /N	Roof integrity	Adaptive grades
9	C1	540	24	1.0	5	1.5	55	IV
10	C1	400	24	1.2	4	1.5	45	IV
11	2+3#	316	6	1.9	5	2.5	85	I
12	B4-upper	383	25	1.2	7	0.6	55	IV
13	B4-upper	433	25	1.2	7	0.6	53	IV
14	B3+4	442	5	2.4	5	2.5	53	IV
15	B4-upper	400	7	0.7	5	0.7	55	II
16	B4-upper	324	8	0.7	5	0.5	55	II
17	2+3#	443	13	2.6	5	2.5	85	II
18	1#	535	10	0.7	5	2.5	55	II
19	B3+4	300	12	3.4	5	2.5	53	II
20	B4-upper	460	8	0.8	7	0.5	55	I
21	2#3#	555	4	1.6	5	2.5	85	I
22	2#	490	8	1.3	5	5	85	II
23	C20	220	15	0.8	5	1	55	II
24	C19	300	17	0.9	4	0.5	53	III
25	C19	320	16	0.7	5	1	53	II
26	C19	300	17	0.9	5	3	53	II
27	C19	330	15	1.1	5	0.9	55	II
28	C24	350	16	1.1	4	3	58	III
29	C20	325	15	1.0	4	3	50	II
30	C24	280	16	1.0	4	0.2	58	III
31	C19	325	15	1.1	4	0.9	55	II
32	C20	360	17	0.7	4	1.1	50	III

Adaptability evaluation results of gob-side entry retaning for FuRong company.

Table 4 Adaptability evaluation results of gob-side entry retaning for DaZhu company

Gateway No.	Name of coal seams	Buried depth /m	Dip angle /(°)	Thickness /m	Lithology (f)	TICIR /N	Roof integrity	Adaptive grades
1	K13	370	37	0.55	2	1.5	78	II
2	K13	670	28	0.83	2	1.9	82	IV
3	K7	200	12	0.83	2	1.8	85	I

Attachment Evaluationresults of adaptability of gob-side entry retaining · 147 ·

Continued Table 4

Gateway No.	Name of coal seams	Buried depth /m	Dip angle /(°)	Thickness /m	Lithology (f)	TICIR /N	Roof integrity	Adaptive grades
4	K13	395	38	0.65	5	4.6	74	II
5	K7	280	12	0.80	4	1.8	75	II
6	K26	270	29	1.47	4.5	1.2	80	II
7	K26	220	28	1.47	4	1.2	80	II
8	K26	460	29	1.47	4	1.4	55	IV
9	K24	544	28	0.60	5	10.2	85	I
10	K24	485	19	0.88	5	4.6	45	II
11	K24	622	21	0.89	5	4.5	80	II
12	K24	540	26	0.46	4	21.0	50	IV
13	K24	490	19	0.88	4	4.6	50	IV
14	Inner	567	19	0.80	5	8.1	70	II
15	Inner	525	19	1.80	9	11.1	70	III
16	Inner	535	14	1.16	5	0.9	60	II
17	Inner	452	30	1.20	4	1.1	70	II
18	Outer	423	15	0.75	8	2.1	70	III
19	Inner	495	30	0.78	2.5	1.0	75	II
20	Inner	350	50	0.88	2.5	0.9	75	III
21	Inner	510	50	0.88	2.5	0.9	75	III
22	Outer	450	25	1.05	4	1.9	85	IV
23	Inner and Outer	560	17	1.58	4	2.1	75	III
24	Inner and Outer	697	20	1.45	5	2.8	70	III
25	26#	442	28	2.87	4.5	2.8	55	IV
26	26#	372	41	1.87	4.5	4.3	55	V
27	21#	344	35	1.31	5	2.9	65	II
28	21#	300	37	1.31	5	2.9	65	II
29	21#	572	47	1.31	5	3.1	65	III
30	21#	669	60	1.35	4.5	4.2	65	II
31	22#	597	32	0.67	7	2.7	85	I
32	21#	534	4	0.46	4.5	2.7	85	I
33	22#	536	25	0.65	5	3.3	65	IV
34	22#	380	30	0.40	5	8.7	73	I
35	22#	622	20	0.80	4.5	0.8	63	I
36	21#	547	33	0.50	6	1.3	85	I
37	21#	545	32	0.92	6	0.6	85	III

Adaptability evaluation results of gob-side entry retaning for DaZhu company.

· 148 ·   Attachment   Evaluationresults of adaptability of gob-side entry retaining

Table 5   Adaptability evaluation results of gob-side entry retaning for GuangWang company

Gateway No.	Name of coal seams	Buried depth/m	Dip angle /(°)	Thickness /m	Lithology (f)	TICIR /N	Roof integrity	Adaptive grades
1	13#	194	53	2.82	6	1.4	65	III
2	13#	178	52	1.85	4	0.4	65	III
3	1#	558	45	2.8	5	0.7	70	IV
4	1#	470	45	2.8	5	1.8	70	IV
5	1#	410	37	2.4	5	1.5	65	III
6	18#	322	28	0.75	5	2.7	65	II
7	18#	384	28	0.75	5	2.7	65	III
8	7#	492	55	0.53	6	2.4	70	II
9	7#	637	52	0.93	5	1.4	75	II
10	7#	575	52	0.93	3	0.3	71	II
11	9#	640	53	0.79	5	8.5	56	II
12	9#	642	51	0.83	5	6.0	73	II
13	12#	307	64	1.17	4	1.2	72	II
14	12#	371	55	0.95	4	0.5	73	II
15	12#	370	64	0.99	4	2.2	72	II
16	11#	637	55	0.82	3	1.5	53	III
17	12#	437	63	0.99	4	1.4	55	II
18	5#	400	40	0.53	6	6.0	65	I
19	5#	470	40	0.56	6	5.7	65	I
20	5#	410	41	1.75	6	1.8	50	IV
21	8#	430	41	0.99	3	1.6	60	III
22	9#-1	450	39	0.84	3	0.6	60	III
23	9#-1	440	41	0.85	3	0.6	70	III
24	9#-1	362	41	0.85	3	5.9	70	III
25	12#-2	418	38	0.52	3.5	3.8	80	II
26	13#	400	42	1.8	4	1.7	80	IV

Adaptability evaluation results of gob-side entry retaning for GuangWang company.